图书在版编目（CIP）数据

秦岭长青：野生生命的庇护所／赵纳勋主编．—北京：北京大学出版社，2013.10

（自然生态保护）

ISBN 978-7-301-23098-5

Ⅰ.①秦… Ⅱ.①赵… Ⅲ.①秦岭－野生植物－植物保护－陕西省 ②秦岭－野生动物－动物保护－陕西省 Ⅳ.①Q948.524.1②Q958.524.1

中国版本图书馆CIP数据核字(2013)第203018号

书　　　　名：	秦岭长青：野生生命的庇护所
著作责任者：	赵纳勋　主编
责 任 编 辑：	黄　炜
标 准 书 号：	ISBN 978-7-301-23098-5/Q·0142
出 版 发 行：	北京大学出版社
地　　　　址：	北京市海淀区成府路205号　100871
网　　　　址：	http://www.pup.cn　　新浪官方微博：@北京大学出版社
电 子 信 箱：	zpup@pup.pku.edu.cn
电　　　　话：	邮购部 62752015　发行部 62750672　编辑部 62752038　出版部 62754962
印 　刷 　者：	北京雅昌彩色印刷有限公司
经 　销 　者：	新华书店
	965毫米×635毫米　16开本　16.25印张　140千字
	2013年10月第1版　2013年10月第1次印刷
定　　　　价：	208.00元

未经许可，不得以任何方式复制或抄袭本书之部分或全部内容。

版权所有，侵权必究

举报电话：（010）62752024　电子信箱：fd@pup.pku.edu.cn

"山水自然丛书"第一辑

"自然生态保护"编委会

顾问

许智宏

主编

吕　植

编委（以姓氏拼音为序）

李　晟　李晟之　苏彦捷　孙　姗　王大军　杨方义
姚锦仙　张树学　赵　昂　赵纳勋　朱小健

摄影／胡万新
Photography by Hu Wanxin

野生生命的庇护所
CHANGQING, QINLING
A Natural Shelter to Wildlife

序一

Preface One

在人类文明的历史长河中，人类与自然在相当长的时期内一直保持着和谐相处的关系，懂得有节制地从自然界获取资源，"竭泽而渔，岂不获得？而明年无鱼；焚薮而田，岂不获得？而明年无兽。"说的也是这个道理。但自工业文明以来，随着科学技术的发展，人类在满足自己无节制的需要的同时，对自然的影响也越来越大，副作用亦日益明显：热带雨林大量消失，生物多样性锐减，臭氧层遭到破坏，极端恶劣天气开始频繁出现……印度圣雄甘地曾说过，"地球所提供的足以满足每个人的需要，但不足以填满每个人的欲望"。在这个人类已生存数百万年的地球上，人类还能生存多长时间，很大程度上取决于人类自身的行为。人类只有一个地球，与自然的和谐相处是人类能够在地球上持续繁衍下去的唯一途径。

在我国近几十年的现代化建设进程中，国力得到了增强，社会财富得到大量的积累，人民的生活水平得到了极大的提高，但同时也出现了严重的生态问题，水土流失严重、土地荒漠化、草场退化、森林减少、水资源短缺、生物多样性减少、环境污染已成为影响健康和生活的重要因素等等。要让我国现代化建设走上可持续发展之路，必须建立现代意义上的自然观，建立人与自然和谐相处、协调发展的生态关系。党和政府已充分意识到这一点，在党的十七大上，第一次将生态文明建设作为一项战略任务明确地提了出来；在党的十八大报告中，首次对生态文明进行单篇论述，提出建设生态文明，是关系人民福祉、关乎民族未来的长远大计。必须树立尊重自然、顺应自然、保护自然的生态文明理念，把生态文明建设放在突出地位，以实现中华民族的永续发展。

国家出版基金支持的"自然生态保护"出版项目也顺应了这一时代潮流，充分体现了科学界和出版界高度的社会责任感和使命感。他们通过自己的努力献给广大读者这样一套优秀的科学作品，介绍了大量生态保护的成果和经验，展现了科学工作者常年在野外艰苦努力，与国内外各行业专家联合，在保护我国环境和生物多样性方面所做的大量卓有成效的工作。当这套饱含他们辛勤劳动成果的丛书即将面世之际，非常高兴能为此丛书作序，期望以这套丛书为起始，能引导社会各界更加关心环境问题，关心生物多样性的保护，关心生态文明的建设，也期望能有更多的生态保护的成果问世，并通过大家共同的努力，"给子孙后代留下天蓝、地绿、水净的美好家园。"

2013年8月于燕园

序二

Preface Two

　　1985年，因为一个偶然的机遇，我加入了自然保护的行列，和我的研究生导师潘文石老师一起到秦岭南坡（当时为长青林业局的辖区）进行熊猫自然历史的研究，探讨从历史到现在，秦岭的人类活动与大熊猫的生存之间的关系，以及人与熊猫共存的可能。在之后的30多年间，我国的社会和经济经历了突飞猛进的变化，其中最令人瞩目的是经济的持续高速增长和人民生活水平的迅速提高，中国已经成为世界第二大经济实体。然而，发展令自然和我们生存的环境付出了惨重的代价：空气、水、土遭受污染，野生生物因家园丧失而绝灭。对此，我亦有亲身的经历：进入90年代以后，木材市场的开放令采伐进入了无序状态，长青林区成片的森林被剃了光头，林下的竹林也被一并砍除，熊猫的生存环境遭到极度破坏。作为和熊猫共同生活了多年的研究者，我们无法对此视而不见。潘老师和研究团队四处呼吁，最终得到了国家领导人和政府部门的支持。长青的采伐停止了，林业局经过转产，于1994年建立了长青自然保护区，熊猫得到了保护。

　　然而，拯救大熊猫，留住正在消失的自然，不可能都用这样的方式，我们必须要有更加系统的解决方案。令人欣慰的是，在过去的30年中，公众和政府环境问题的意识日益增强，关乎自然保护的研究、实践、政策和投资都在逐年增加，越来越多的对自然充满热忱、志同道合的人们陆续加入到保护的队伍中来，国内外的专家、学者和行动者开始协作，致力于中国的生物多样性的保护。

　　我们的工作也从保护单一物种熊猫扩展到了保护雪豹、西藏棕熊、普氏

摄影／赵纳勋
Photography by Zhao Naxun

原羚，以及西南山地和青藏高原的生态系统，从生态学研究，扩展到了科学与社会经济以及文化传统的交叉，及至对实践和有效保护模式的探索。而在长青，昔日的采伐迹地如今已经变得郁郁葱葱，山林恢复了生机，熊猫、朱鹮、金丝猴和羚牛自由徜徉，那里又变成了野性的天堂。

然而，局部的改善并没有扭转人类发展与自然保护之间的根本冲突。华南虎、白暨豚已经趋于灭绝；长江淡水生态系统、内蒙古草原、青藏高原冰川……一个又一个生态系统告急，生态危机直接威胁到了人们生存的安全，生存还是毁灭？已不是妄言。

人类需要正视我们自己的行为后果，并且拿出有效的保护方案和行动，这不仅需要科学研究作为依据，而且需要在地的实践来验证。要做到这一点，不仅需要多学科学者的合作，以及科学家和实践者、政府与民间的共同努力，也需要借鉴其他国家的得失，这对后发展的中国尤为重要。我们急需成功而有效的保护经验。

这套"自然生态保护"系列图书就是基于这样的需求出炉的。在这套书中，我们邀请了身边在一线工作的研究者和实践者们展示过去30多年间各自在自然保护领域中值得介绍的实践案例和研究工作，从中窥见我国自然保护的成就和存在的问题，以为热爱自然和从事保护自然的各界人士借鉴。这套图书不仅得到国家出版基金的鼎力支持，而且还是"十二五"国家重点图书出版规划项目——"山水自然丛书"的重要组成部分。我们希望这套书所讲述的实例能反映出我们这些年所做出的努力，也希望它能激发更多人对自然保护的兴趣，鼓励他们投入到保护的事业中来。

我们仍然在探索的道路上行进。自然保护不仅仅是几个科学家和保护从业者的责任，保护目标的实现要靠全社会的努力参与，从最草根的乡村到城市青年和科技工作者，从社会精英阶层到拥有决策权的人，我们每个人的生存都须臾不可离开自然的给予，因而保护也就成为每个人的义务。

留住美好自然，让我们一起努力！

吕植

2013 年 8 月

前言

Foreword

位于中国中部的秦岭山脉，不仅是长江、黄河两大水系的分水岭、南北气候的分界线，也是动物区系东洋界和古北界的交汇地。独特的地理位置，良好的森林植被，孕育着丰富的野生动植物资源，而长青则是秦岭野生生命分布的精华之所在。

作为秦岭的自然保护工作者，我们目睹了长青由于森工企业采伐，原始森林消失、大熊猫栖息地萎缩、种群数量下降的过程。我们也见证了北京大学潘文石教授带领的研究小组在极其艰苦和困难的情况下进驻长青，开始长达十余年的秦岭大熊猫野外研究，并为保护这一区域的野生大熊猫及其栖息地与他的团队奔走呼吁，致信国家领导人，使长青从一个森工企业的采伐地成功转型为自然保护区。我们也欣喜地看到，由于GEF（全球环境基金）项目在长青保护区的实施，使曾经从事林木采伐的职工很快踏上了保护之路。经过多年的细心呵护，我们感受到了长青保护区日新月异的变化，森林在更新，栖息地在恢复，大熊猫及同域分布的动物在不断增加，生物多样性愈加丰富，昔日的采伐迹地和林区采伐道路成为野生动物活动的场所。长青已成为野生动物栖息繁衍的乐土，人与自然和谐的家园。

在从事野生动物的保护中，在野外巡护的山林里，在科研监测的样线上，我们经常与大熊猫等珍稀动物零距离接触，与羚牛等珍禽异兽近距离对峙，与朱鹮等大自然的精灵相伴于同一片天地。长青动植物灵动的身影、曼妙的舞姿、绚丽的色彩，以及它们的优良的栖息地吸引了我们，触动了我们，也感动了我们，促使我们用手中的相机在工作中记录下一幅幅动人的画面和一个个美丽的

摄影/赵纳勋
Photography by Zhao Naxun

身影。对大自然的热爱和保护之路任重道远，我们感觉到有义务和责任将它们展示给同样热爱自然、热爱大熊猫、热爱绿色和热爱一切野生生命的朋友。

在本书的编辑出版过程中，得到了北京大学王大军博士和李晟博士、陕西师范大学任毅教授、中科院成都生物所李成博士和重庆野趣科技有限公司张巍巍先生的支持和帮助，在此表示衷心的感谢。由于时间仓促、水平有限，错误在所难免，敬请批评指正。

赵纳勋

Situated in the middle of China, the Qinling Mountains stand not only as the divide between the Changjiang River and the Yellow River systems, the demarcation line between the climates of the North and the South, but also a converging place for the fauna of both the Oriental realm and the Palaearctic realm. Its unique geography

野生生命的庇护所
CHANGQING, QINLING
A Natural Shelter to Wildlife

and excellent forestation nurture rich varieties of wild plants and animals, and Changqing, in particular, provides the most remarkable habitat for the wild life of the Qinling Mountains.

During my years of work in the Qinling Mountains as a conservation worker, I saw the whole process of deforestation, the shrinking of the wildlife habitat, and the dwindling of the giant panda population due to heavy commercial logging. I also luckily witnessed the establishment of the nature reserve in this region with a view to protecting giant pandas at the request of Professor Pan Wenshi of Peking University and his research group, who spent more than ten years in Changqing doing field research on giant panda in the Qinling Mountains despite the hardships of their living condition. We give them full credit for their indefatigable effort to keep urging people and petitioning to government leaders to protect the wild giant panda population and their habitat. It is largely thanks to them that we now see the transformation of Changqing from a timber land to a nature reserve. I am delighted to see that the previous timber workers are converting to conservation staff, thanks to the projects funded by GEF (Global Environmental Fund). Through years of work in conservation, I am more than happy to see the ever-changing achievements of the Changqing nature reserve: the forest is coming back, the habitat is recovering, the populations of giant panda and other wild animals are on the increase, the biodiversity is getting richer, the logged land and the abandoned timber trails are becoming the playgrounds of the wildlife. The Changqing nature reserve has turned into a paradise for both wildlife and humans, and it represents a great harmony between man and nature.

During our years of work in conservation, by patrolling along forest trails and monitoring on the designed transects, we often make close contacts with such precious animals as giant pandas, takins and crested ibis. We enjoy sharing the same world with them and we have often been touched by such wonderful scenery and events in nature. Their elegant figures, graceful dances, brilliant colors, and their easy life in the habitat inspired us to capture the moments of beautiful life in the wild in Changqing. However, there is still a long way to go on our mission in the name of love for and protection of nature, and we hope that, by presenting this volume to the readers, we will be joined by more who love nature, care giant panda, think green and take wildlife as life.

In compiling and publishing this book, we are particularly grateful for the great support and help of Dr. Wang Dajun and Dr. Li Sheng of Peking University, Professor Ren Yi of Shaanxi Normal University, Dr. Li Cheng of Chengdu Institute of Biology, Chinese Academy of Sciences, and Mr. Zhang Weiwei from Chongqing Wild Life's Technology Co, Ltd. Comments and suggestions are always welcomed to help us improve this volume.

Zhao Naxun

摄影 / 赵纳勋
Photography by Zhao Naxun

野生生命的庇护所
CHANGQING, QINLING
A Natural Shelter to Wildlife

目录

Contents

一、从森工采伐地到自然保护区 From Forestry Logging Area to Nature Reserve	1
二、长青的景致 The Scenery at Changqing	11
春 Spring	14
夏 Summer	18
秋 Autumn	24
冬 Winter	30
三、长青的四大国宝——大熊猫、金丝猴、羚牛、朱鹮 The Four National Treasures in Changqing: Giant Panda, Golden Monkey, Takin and Crested Ibis	35
秦岭大熊猫 The Qinling Giant Panda	35
秦岭金丝猴 The Qinling Golden Monkey	47
秦岭羚牛 The Qinling Takin	55
朱鹮 Crested Ibis	63

四、长青的动物大家庭 Changqing, Home to the Animals	73
兽类 Mammals	73
两栖爬行动物 Amphibians & Reptiles	83
鱼类 Fish	93
鸟类 Birds	97
昆虫类 Insects	175
五、长青的奇葩 The Wonders of Changqing	189
植被垂直分布 The Vertical Distribution of Vegetation	190
时间的花序 The Inflorescence of Time	192
六、人与自然和谐的家园 Human and Nature in Harmony	227
七、索引 Index	236

秦岭长青

Changqing, Qinling
A Natural Shelter to Wildlife

一

从森工采伐地到自然保护区

From Forestry Logging Area to Nature Reserve

位于陕西省境内的秦岭中段山脉，是中国大熊猫最北的家园，是野生大熊猫分布密度最高的地区，是濒临灭绝的朱鹮发现地。以大熊猫等珍稀野生动植物为主要保护对象的长青国家级自然保护区，位于秦岭南坡洋县境内，面积299.06平方千米，是在原森工企业长青林业局停止采伐、转产分流的基础上建立的。其总体目标是：永久性地保护和维持长青自然保护区的整体生物多样性，特别是维持能在特殊环境中生存的大熊猫种群，并恢复遭到破坏的生态系统。

长青林业局是由国家计委于1967年批准建立的，"长青"之名源于"青山常在，永续利用"的林业经营理念。鼎盛时期职工达2262人，拥有完备的采运、营林、筑路、木材综合利用等多种经营生产体系和管理机构。经营区面积138255公顷，年均生产木材3万立方米，累计生产木材78万立方米，共消耗森林蓄积172万立方米，为国家经济建设做出了重要贡献。但20多年的高强度采伐使区内森林资源日益枯竭，生物多样性遭到严重破坏，大熊猫栖息地安全受到严重威胁。为了便于经营区管理和加强自然保护工作，部分区域于1979年被划出，分别建立了龙草坪林业局和佛坪国家级自然保护区。1992年林业部将长青林业局纳入了大熊猫保护工程总体规划，长青林业局在采伐的同时也开展了一系列大熊猫保护与抢救工作，为后续的自然保护工作奠定了基础。

1985年，长青林业局与北京大学潘文石教授领导的大熊猫研究小组合作，开展了长达十余年的大熊猫生态生物学研究，奠定了北京大学在大熊猫野外研究方面的重要地位，也使正在采育的森工企业步入了保护之路，培养了一批致力于野生动物保护和科学研究的专家学者。

摄影／赵纳勋
Photography by Zhao Naxun

　　面对林区过度采伐、林权林界不清以及社区林副业生产等活动对大熊猫栖息地的干扰和破坏，1993年，潘文石及他的研究团队致信国家领导人、国务院和全国人大环境资源委员会，陈述了秦岭正在发生的生态危机，呼吁长青林业局转产，尽快建立大熊猫保护区。他们的呼吁引起了国家领导人和政府的高度重视。1994年，长青林业局转产开始，并停止了采伐，同年12月陕西省政府批准建立省级长青自然保护区。1995年12月国务院以国函129号正式批复"同意建立长青国家级自然保护区"。

　　长青保护区地处我国南北气候的分界线和动植物区系的交汇过渡地带，区内最高海拔3071米，最低海拔700米，森林覆盖率达97%以上。独特的地理位置和气候条件孕育了区内丰富的动植物多样性。辖区内已知有高等植物

2100 余种（种子植物 1556 种，蕨类植物 61 种，苔藓类植物 400 余种），其中一级保护植物 2 种，二级保护植物 10 种。野生脊椎动物 29 目 78 科 213 属 435 种，其中兽类 63 种，两栖类和爬行类 35 种，鸟类 319 种，鱼类 18 种。国家一级重点保护野生动物有大熊猫、金丝猴、羚牛、林麝、豹、朱鹮、金雕等 9 种；二级重点保护野生动物有黑熊、斑羚、红腹角雉、大鲵等 52 种。

The middle stretch of the Qinling Mountains within Shaanxi province is the northernmost home to the densest populations of giant pandas. It is also the location where the endangered crested ibis was first discovered. Located on the southern slopes of Qinling with an area of 299.06 square kilometers, the Changqing National Nature Reserve shelters numerous unique species of plants and animals including the giant panda. Transformed from its previous role as a logging enterprise, Changqing Forestry Bureau diverted most of its resources to establishing the current nature reserve. The objectives of the nature reserve are to permanently conserve the biodiversity within the limits of the nature reserve, with a special aim to maintain the giant panda population on the nature reserve, and to rehabilitate local ecosystem that has been damaged by human activity.

Changqing Forestry Bureau was approved for establishment by the National Planning Commission in 1967. The name Changqing, meaning evergreen, is derived from the management philosophy of the Bureau, reflecting its idea of sustainable development and utilization of the natural resources. At the peak of its operation, the bureau comprised of 2262 employees and capacitated logging, delivery, forest management, road construction, timber application, diversified production and management mechanisms. The operation area reached 138,255 hectares with an annual timber output of 30,000 cubic meters and a cumulated output of 780,000 cubic meters of lumber. In total, 1,720,000 cubic meters of forest was consumed, which contributed significantly to the economic development of the nation. However, over 20 years of intensive logging progressively depleted the forest's resources, severely devastated biodiversity and threatened panda habitats within Changqing. In order to facilitate forestry management and enhance conservation, part of the forests was marked out for restoration in 1979, followed by the establishment of Longcaoping Forestry Bureau and the Foping National Nature Reserve. In 1992, Changqing Forestry Bureau was incorporated by the National Forestry Department into the overall conservation project planning for the giant panda, thus engaging itself in a series of conservation projects and panda rescue missions while maintaining a moderate level of logging. This laid a solid foundation for the nature reserve's future role in conservation.

In 1985, Changqing Forestry Bureau began co-operation of Peking University's Professor Pan Wenshi and his giant

摄影／向定乾
Photography by Xiang Dingqian

摄影／向定乾
Photography by Xiang Dingqian

秦岭长青 野生生命的庇护所
CHANGQING, QINLING
A Natural Shelter to Wildlife

panda research team, conducting over 10 years of ecological and biological research on the giant panda. This effort established Peking University's leading role in field research work on giant pandas and paved the way for the conservation of the local logging industry. In the meantime, it fostered a group of experts and scholars who are dedicated to wildlife conservation and research.

In 1993, facing the threat to the panda population caused by deforestation, ambiguous land boundaries or ownership, and auxiliary agricultural activities in the forest, Professor Pan Wenshi and his research team wrote a petition letter to the central government leaders, the State Council and the National People's Congress Environment and Resources Committee. The group reported the urgent ecological crisis in Qinling, advocated the need for the forestry's reassignment, and called for the immediate establishment of nature reserves to protect the giant panda in Qinling. The appeal drew grave attention from government leaders. In 1994, Changqing Forestry Bureau started its reconfiguration and logging was called to a halt. In December of the same year, the Shaanxi Provincial Government approved the bill for establishing the Changqing Provincial Nature Reserve. In December of 1995, the State Council issued a reply letter No.129, officially approving "Consent to the Establishment of the Changqing National Nature Reserve".

The Changqing Nature Reserve is a transitional region that sits cross the climatic divide between North and South China, and it traverses an area where many zoological and botanical regions of the country overlap. With the highest elevation reaching 3,071 meters and lowest elevation dropping below 700 meters, the reserve has a vegetation coverage of over 97%. The area's unique geographic location and climate conditions have nurtured a rich diversity of plant and animal species. Discovered wildlife include over 2,100 species (1,556 species of seed plants, 61 species of ferns, over 400 species of bryophytes) of higher plants, of which 2 are Class I state protected species and 10 are Class II state protected species. 29 orders (78 families, 213 genera, 435 species) of wild vertebrates including 63 species of mammals, 35 species of reptiles and amphibians, 319 species of birds and 18 species of fish. Among them, 9 species including the wild giant panda, golden monkey, takin, musk deer, panther, crested ibis and golden eagle are Class I protected species. 52 Class II state protected species inhabiting the area include the Asian black bear, the Himalayan goral, Temminck's tragopan and the Chinese giant salamander, etc.

秦岭长青
野生生命的庇护所
CHANGQING, QINLING
A Natural Shelter to Wildlife

摄影／赵纳勋
Photography by Zhao Naxun

摄影／赵纳勋
Photography by Zhao Naxun

二

长青的景致

The Scenery at Changqing

地处秦岭南坡腹地的长青自然保护区，平均海拔1700米，这里森林茂密，河谷众多，水质清纯，气候多变，风光绮丽。区内崇山峻岭与飞瀑流泉相映生辉，原始森林与茫茫竹海交错分布，高山草甸与冰川遗迹天造地设，奇花异草和珍禽异兽增光添彩。仲春，处处山花烂漫，百鸟齐鸣，仿若进入天然花坛；盛夏，四野滴翠，清风徐来，如入天然氧吧；金秋，万山红遍，层林尽染，一如五彩锦缎；严冬，银装素裹，粉堆玉砌，一派异域景象。

Located on the southern slopes of the Qinling Mountains, with an average altitude of 1,700 meters above sea level, the Changqing Nature Reserve has dense forests, numerous rivers, valleys, and pure water resource. The weather is volatile and the landscape is beautiful. The reserve grounds encompass magnificent mountain ranges with surging waterfalls and streams. Primitive wooded forests mingle with bamboo forests. Alpine meadows and remaining glaciers meet in natural harmony. Unique flora and fauna garner further spotlight to the area. In spring, flower petals cover all the hills and slopes which become a naturally formed flower bed where birds of every kind chirp in unison. Midsummer renders all the fields emerald green, with breezes steadily ringing in with fresh air, making the grounds a natural oxygen bar. Autumn is golden, with mountain slopes covered layer by layer in red and orange like colorful brocade bands. In winter, the snow covers the landscape with silvery white as if piled with jade, likening the landscape to that of exotic foreign lands.

美丽长青：
春、夏、秋、冬

Splendid Changqing:
Spring, Summer, Autumn and Winter

春 S P R I N G

百米高的石门瀑布仿佛从天而降，瞬间化作漫天烟云，亦真亦幻，如歌似画！摄影／赵纳勋
The hundred-meter high Shimen Falls appear to cascade from the sky. In a quick instant, falling water dissolves into frolicking vapors that fade away like a mirage. Photography by Zhao Naxun

1. 阳春三月，春风化雨的同时，亦会化作漫天大雪，在长青山岭间形成雪压花枝俏的美丽景象。摄影／向定乾
As it gently gets warm in March, breezes nurture rains, sometimes bringing forth spring flurries, gently covering the mountain ridges with delightful snow-capped flowers. Photography by Xiang Dingqian

野生生命的庇护所
CHANGQING, QINLING
A Natural Shelter to Wildlife

2,3.长青的春天,千树琼花,绿满峰岭,仿佛一夜之间苏醒的少女,娇媚可人,风情无限。
摄影/赵纳勋
Springtime in Changqing, with thousands of viburnums, and green-filled peaks, is like a maiden waking from her slumber, delightful and enchanting. Photography by Zhao Naxun

4.杜鹃开得姹紫嫣红,汪洋恣肆。杜鹃花开的季节,万绿丛中处处红,长青无山不飞花……
摄影/赵纳勋
Rhododendrons bloom in rich violets and reds. During this season, the green mountains are dotted here and there with red. Every mountain in Changqing is showered with flower petals. Photography by Zhao Naxun

夏

S U M M E R

溪瀑溅落如银,飞散似花,如轻纱曼舞,又似银链抖落。摄影/赵纳勋
Small creeks come splashing down from the rocks, tossing white petals of foam in the sky, like droplets of silver cascading from the sky. Photography by Zhao Naxun

1. 初夏的长青，苍山黛石之间，芳草萋萋，山花竞开，俨然一幅如诗似画的醉人美景。摄影／赵纳勋
Early summer in Changqing. Amidst the umber rocks in the green mountains are lush grass and wild flowers in full bloom, presenting a poetic and picturesque leadscape. Photography by Zhao Naxun

2. 久雨初晴，山岚如烟，岩石在轻云流雾的遮掩中，影影绰绰，忽隐忽现。摄影／赵纳勋
Sunshine after a long spell of rain. As it clears up, the mountains loom from behind the misty veil and the rocks peep out from under the shadow and fickle canopy of fleeting clouds. They veil the rocky slopes, blurring their shapes one moment and exposing them another. Photography by Zhao Naxun

3. 山巅之上，重岭之间，云雾缭绕，胜似仙境。摄影／赵纳勋
Upon the pinnacles and between the ranges, clouds permeate the forest, transforming it into a dreamland. Photography by Zhao Naxun

1	2
	3

野生生命的庇护所
CHANGQING, QINLING
A Natural Shelter to Wildlife

1. 第一缕晨曦穿透松林，高山草甸沐浴在朝阳的清辉中。摄影／赵纳勋
When the first ray pierces through the pine trees, the meadow wakes to the glory of the morning sun. Photography by Zhao Naxun

2. 万山起伏，苍茫如黛。在离蓝天最近的地方，草甸尽情地一展天然容颜。摄影／赵纳勋
Miles upon miles of mountains rise and fall into an ink-colored vastness. Where the land is closest to the sky, the alpine meadow freely displays its natural poise. Photography by Zhao Naxun

3. 朝霞万丈，云海茫茫，长青的每个清晨都如画如诗。摄影／赵纳勋
At sunrise, the golden morning glow expands over a sea of clouds. Every morning in Changqing is enchanting. Photography by Zhao Naxun

4. 飞瀑流泉潇潇洒洒高歌出山，瀑落珠玑，水溅成花。摄影／赵纳勋
The boisterous waterfall surges carelessly down the mountain, orchestrating a song on its way by shooting off gem-like beads and splashing flowery sprinkles. Photography by Zhao Naxun

AUTUMN

野生生命的庇护所
Changqing, Qinling
A Natural Shelter to Wildlife

秋风将树叶吹成了绚烂的彩蝶,只有秋水依然端庄,清清浅浅,不嬉不闹。摄影/赵纳勋
The autumn wind brushes the leaves with luscious colors, like those of butterflies. Only the water remains solemn as before, clear and brisk, without a sign of mischief. Photography by Zhao Naxun

1. 山腰是黄绿相间的森林，山顶是飘飘忽忽的云雾，秋天的长青，俨然一个五彩斑斓的大花园。摄影／向定乾
Golden and green forests cover the mountain slopes, while drifting clouds hover over the peaks. Autumn in Changqing brings to mind a colorful garden. Photography by Xiang Dingqian

2. 老树古枝擎天，写意舒卷，叶是璀璨的金黄，枝是斑驳的沧桑。摄影／向定乾
With aged branches extending skywards, wavering with lighthearted whims, old trees display the brilliance of their golden leaves and age-worn speckled barks. Photography by Xiang Dingqian

3. 秋风为群山抹上彩妆。在激越的秋声中，彩色的树叶挥洒着一季的灿烂心情。摄影／赵纳勋
As the wind adorns the mountains with new colors, the colorful leaves dance to the rhythm of autumn, as if celebrating the joy of the season. Photography by Zhao Naxun

4. 那些树，那片云，那静静的山谷和所有关于风的语言，都与生命、时间和永恒有关……摄影／赵纳勋
The woods, the clouds, the quiet valley and all the murmuring of the winds, they all tell of life, time, and eternity. Photography by Zhao Naxun

5. 那一刻万籁俱寂，只有隐约的风声。于斑斓的深处，黄叶飒飒，林涛阵阵，树木醉人的金黄渲染了群山的深秋。摄影／赵纳勋
When all is silent, only the subtle whistles of the wind can be heard. In the depths of the thick colors, yellow leaves rustle. Golden hues sweep across the mountains in the autumn air. Photography by Zhao Naxun

野生生命的庇护所
CHANGQING, QINLING
A Natural Shelter to Wildlife

1 | 3
2

1. 依然喧响的瀑声，风中翻飞的彩蝶，共同奏响了大山深处沉郁苍劲的秋天交响乐。摄影／赵纳勋
Bellowing water-falls and fluttering butterflies come together in the deep mountains to create majestic symphonic music for autumn. Photography by Zhao Naxun

2. 金黄的树叶染红了溪流里的石头，连水也变成了彩色。只有长青的山谷才会有如此神奇的变化。摄影／赵纳勋
The golden foliage has dyed the river rocks red, coloring even the flowing water. Such magic only takes place in Changqing. Photography by Zhao Naxun

3. 一抹青绿，一抹铭黄，长青在季节的边缘变幻着四季的色调，尽情抒写着生命的七彩诗篇。摄影／赵纳勋
A layer of green and a layer of amber, the seasonal colors of Changqing merge at the borderlines, passionately proclaiming the psalms of life. Photography by Zhao Naxun

野生生命的庇护所
CHANGQING, QINLING
A Natural Shelter to Wildlife

WINTER

野生生命的庇护所
CHANGQING, QINLING
A Natural Shelter to Wildlife

高贵的朱鹮，不畏严寒。摄影／张永文
Distinguished crested ibises defy severe cold.
Photography by Zhang Yongwen

1. 风雪中的常绿植物。摄影／赵纳勋
Evergreen trees in the snow. Photography by Zhao Naxun

2. 雪约霜期，万山洗尽铅黄。蓝天下，树的剪影全变成了一副冰肌玉骨。摄影／胡万新
When frost falls, all the mountains seem to be cleaned of their mottled colors. Against the blue sky, the trees become ice figurines. Photography by Hu Wanxin

3. 长青储满了冬天的心绪，将厚厚的清冷化成了心灵的守候。摄影／张永文
At this moment, Changqing is brimming with wintry sensibilities, ready to transform the profound quietude and inactivity of the season into a time of spiritual waiting. Photography by Zhang Yongwen

4. 一场冬雪铺天盖地，长青变成了一幅静美的水墨长幅……摄影／赵纳勋
An all-pervading snow transforms Changqing into a serene Chinese watercolor. Photography by Zhao Naxun

5. 摄影／张永文
Photography by Zhang Yongwen

1	
2	4
3	5

三

长青的四大国宝
大熊猫、金丝猴、羚牛、朱鹮

The Four National Treasures
in Changqing:
Giant Panda, Golden Monkey,
Takin and Crested Ibis

秦岭大熊猫
The Qinling Giant Panda

  1962 年，郑光美院士在秦岭南坡发现大熊猫标本，并经实地考察后，于 1964 年科学报道了秦岭大熊猫的存在，1986 年在洋县金水河口发现的大熊猫化石证明：大熊猫在秦岭的存在有着悠久的历史，秦岭南坡是秦岭大熊猫的故乡。2005 年，秦岭的大熊猫被学术界确定为"大熊猫秦岭亚种"。大熊猫是长青保护区的主要保护对象，其数量约占秦岭大熊猫种群的三分之一。保护区的兴隆岭及其周边地区是秦岭大熊猫分布的核心区域。

  In 1962, Professor Zheng Guangmei found a giant panda specimen on the southern slope of the Qinling Mountains. Research was carried out accordingly and the proof of the existence of living wild pandas was published in 1964. The discovery of panda fossil in 1986, at Jinshui River in Yangxian County, proved the extensive period pandas had been inhabiting Qinling, and that the southern slopes of the mountains were the original home to Qinling giant pandas. In 2005, Qinling giant pandas were classified as the Qinling sub-species. These pandas, taking up 1/3 of the total population of giant pandas in Qinling area, are the main target of conservation at the Changqing National Nature Reserve. Xinglongling Peak within the reserve grounds and its neighboring areas are the core area of Qinling panda distributions.

摄影 / 赵纳勋
Photography by Zhao Naxun

大熊猫夏居地。摄影 / 胡万新 & 赵纳勋
The giant panda's summer habitat. Photography by Hu wanxin & Zhao Naxun

野生生命的庇护所
CHANGQING, QINLING
A Natural Shelter to Wildlife

大熊猫冬居地。摄影／赵纳勋
The giant panda's winter habitat. Photography by Zhao Naxun

野生生命的庇护所
CHANGQING, QINLING
A Natural Shelter to Wildlife

1. 竹子是它的最爱。摄影/赵纳勋
Bamboo is his favorite food. Photography by Zhao Naxun

2. 最美味的还是竹笋。摄影/赵纳勋
Bamboo shoot is delicious. Photography by Zhao Naxun

3. 偶尔也打个牙祭。摄影/张永文
An occasional meat snack. Photography by Zhang Yongwen

4. 渴了喝口清澈的山泉。摄影/向定乾
Mineral water beats thirst. Photography by Xiang Dingqian

5. 不同的食物，不同的粪便。摄影/赵纳勋
Different diet produces different manure. Photography by Zhao Naxun

		3	4
1	2		5

1. 健康的身体离不开矿物质。摄影／赵纳勋
Obtaining his daily dose of minerals. Photography by Zhao Naxun

2. 树上也是大熊猫交配的场所。摄影／向定乾
Trees: also a good mating location for pandas. Photography by Xiang Dingqian

3. 通过气味寻找异性。摄影／胡万新
Searching for the opposite sex by smell. Photography by Hu Wanxin

4. 春季是大熊猫的繁殖交配期。摄影／向定乾
Spring is the mating season for pandas. Photography by Xiang Dingqian

5. 留下标记划分领地。摄影／向定乾
Marking his territories. Photography by Xiang Dingqian

1. 秦岭大熊猫以岩洞为巢穴。摄影／赵纳勋
Qinling giant panda makes its home in caves. Photography by Zhao Naxun

2. 8~9月是秦岭大熊猫的产仔期。摄影／赵纳勋
August to September is the breeding season for the giant pandas. Photography by Zhao Naxun

3. 没有冬眠的习性。摄影／向定乾
Pandas do not hibernate. Photography by Xiang Dingqian

4. 冬天依然悠闲。摄影／向定乾
Winter is relaxing, too. Photography by Xiang Dingqian

5. 湍急的河流也无所畏惧。摄影／向定乾
Rapids are not a threat to pandas. Photography by Xiang Dingqian

1	3	4
2		5

1. 洞穴是大熊猫幼仔温暖的家。摄影／赵纳勋
Cave is cozy home to panda cub. Photography by Zhao Naxun

2. 曾经的采伐地成为大熊猫的乐园。摄影／赵纳勋
Former logging sites transformed into a haven for giant pandas. Photography by Zhao Naxun

3. 树上可沐浴阳光也可躲避天敌。摄影／赵纳勋
Tree branches provide sunshine and shelter from predators. Photography by Zhao Naxun

秦岭长青 野生生命的庇护所
CHANGQING, QINLING
A Natural Shelter to Wildlife

秦岭金丝猴
The Qinling Golden Monkey

与大熊猫一样，秦岭也是中国金丝猴分布的最北限，目前约有3800～4000只。长青自然保护区是秦岭金丝猴的高密度分布区，在海拔1500～3000米的落叶阔叶林、针阔混交林和亚高山针叶林带，经常可以看到金丝猴在林中跳跃、欢呼奔走的优美身姿。

Qinling is not only the northernmost home to giant pandas, but also to golden monkeys, the latter enjoying a population of 3,800 to 4,000. The Changqing National Nature Reserve is a densely populated area for golden monkeys usually dwelling at an altitude of 1,500 to 3,000 meters within deciduous broadleaf forests, needleleaf-broadleaf forests and subalpine needeleaf forests. Golden monkeys can often be seen in the forests.

摄影／赵纳勋
Photography by Zhao Naxun

1. "哨猴"担负群体的警戒任务。摄影／赵纳勋
Guard monkey is responsible for the security of the group. Photography by Zhao Naxun

2，3. 金丝猴以树栖为主，但也要在地面捡拾果实和喝水。摄影／赵纳勋
Golden monkeys are most arboreal, but from time to time they have to seek food and water from the ground. Photography by Zhao Naxun

4. 金丝猴以群居性家族社会生活。摄影／胡万新
Golden monkeys are gregarious animals. Photography by Hu Wanxin

野生生命的庇护所
CHANGQING, QINLING
A Natural Shelter to Wildlife

| 1 | | 3 | | |
| 2 | | 4 | 5 | |

1,2.树叶和果实是金丝猴的主要食物。摄影／赵纳勋
Leaves and fruits are the staple food for golden monkeys. Photography by Zhao Naxun

3.相互理毛，既是亲昵，又是补充营养物质。摄影／赵纳勋
Mutual grooming brings intimacy and additional nutrition for these monkeys. Photography by Zhao Naxun

4.每年九、十月份是金丝猴的主要婚配期。摄影／赵纳勋
September to October is the mating period for the monkeys. Photography by Zhao Naxun

5.经过7个月的妊娠期，母猴在三、四月份分娩。摄影／赵纳勋
After 7 months of pregnancy, female monkeys give birth in March and April. Photography by Zhao Naxun

野生生命的庇护所
CHANGQING, QINLING
A Natural Shelter to Wildlife

野生生命的庇护所
CHANGQING, QINLING
A Natural Shelter to Wildlife

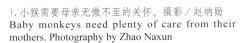

1. 小猴需要母亲无微不至的关怀。摄影／赵纳勋
Baby monkeys need plenty of care from their mothers. Photography by Zhao Naxun

2. 两年生一胎，每胎一般只产一崽。摄影／赵纳勋
One birth every two years, yielding only one infant each time. Photography by Zhao Naxun

3. 金丝猴 5~6 岁才能达到性成熟。摄影／赵纳勋
A golden monkey reaches sexual maturity at age 5 or 6. Photography by Zhao Naxun

4. 雄性体大魁梧，毛色艳丽，嘴角肉状突大而明显。摄影／赵纳勋
Burly male monkeys have a gorgeous coat and thick protrusions mark on the mouth. Photography by Zhao Naxun

秦岭羚牛
The Qinling Takin

秦岭最大的野生动物秦岭羚牛在长青保护区广泛分布，它们以青草、树叶、枝芽和树皮为食，体重可达 400 千克左右，虽是庞然大物却能在悬崖峭壁上奔走自如。随着季节的变化，秦岭羚牛会迁徙到不同的海拔高度栖息。初夏时节，它们聚会于高山草甸，少则几十，多则上百，享受高山的凉爽，寻找心仪的伴侣。秋冬季节，它们家族式活动于中低海拔的针阔混交林和阔叶林带。翌年春季，小宝宝就会降临世间。

The largest wild animals in Qinling are the takins that are widely distributed in the Changqing Nature Reserve. They live on grass, tree leaves and bark, and weigh around 400 kilograms. Despite their large physical sizes, they can move along mountain cliffs with great speed and ease. As the seasons vary, Qinling takins migrate to different sea levels. During early summer, they gather in the alpine meadows in tens to hundreds cooling off and looking for mates. In the fall, whole families migrate to mixed needleleaf-broadleaf forests and broadleaf forests at mid-low sea level. Calves are born the following spring.

摄影／向定乾
Photography by Xiang Dingqian

秦岭长青 野生生命的庇护所
CHANGQING, QINLING
A Natural Shelter to Wildlife

1. 羚牛是秦岭山区最大的野生动物。摄影／赵纳勋
Takin is the largest animal in the Qinling mountains. Photography by Zhao Naxun

2. 秦岭羚牛是素食主义者，森林提供了它们充足的食物。摄影／赵纳勋
Takins are vegetarian and the forest provides them with adequate food. Photography by Zhao Naxun

3，4. 秦岭羚牛喜欢群居生活。摄影／赵纳勋 & 雍严格
Takins are gregarious animals. Photography by Zhao Naxun & Yong Yange

1	3	4
2		5

1.六、七月份是秦岭羚牛繁衍后代的好时机。红外相机摄影
June and July are good months for reproduction. Infrared Photography

2.春季小宝宝就会降临世间。摄影／赵纳勋
Baby calve will soon be born. Photography by Zhao Naxun

3.秋冬季节秦岭羚牛以家族为单位活动。摄影／赵纳勋
Takins live gregariously during the fall and winter. Photography by Zhao Naxun

4.未成年的小羚牛离不开妈妈的照顾。摄影／张希明
A calf cannot live without its mother's care. Photography by Zhang Ximing

5.森林里的童年很快乐。红外相机摄影
Life in the forest is filled with joy. Infrared Photography

秦岭长青 野生生命的庇护所
CHANGQING, QINLING
A Natural Shelter to Wildlife

野生生命的庇护所
CHANGQING, QINLING
A Natural Shelter to Wildlife

1. 秦岭羚牛不惧严寒，但惧怕……摄影／向定乾
Takin is not afraid of the cold, but when it comes to...
Photography by Xiang Dingqian

2. 庞大的身躯在悬崖峭壁上行走自如。摄影／向定乾
Large in size, takins can move about on the cliffs with ease.
Photography by Xiang Dingqian

朱鹮
Crested Ibis

　　1981年，濒临灭绝的朱鹮被科学家发现于秦岭南坡洋县，经过二十多年的就地保护和人工繁殖，其种群数量由7只发展壮大到2000多只，栖息地范围不断扩大，分布越来越广，长青华阳也已成为朱鹮栖息的乐园，随处可见朱鹮翘立枝头、田间觅食、河边嬉戏、空中翱翔的美丽画卷……

　　In 1981, the endangered crested ibis was found in Yangxian County on the southern slope of the Qinling Mountains. Through the conservation efforts for over two decades, the overall population of crested ibis has reached over 2,000 from the initial 7. The habitat of crested ibis has been expanding and their distributions are growing wider. Changqing has become a haven to these birds and they can be seen perching on tree branches, feeding in the paddy fields and frolicking in the creeks or soaring elegantly in the sky…

摄影／赵纳勋
Photography by Zhao Naxun

1. 每年的三、四月份是朱鹮的婚配期。摄影／赵纳勋
March to April is the mating season of the crested ibis. Photography by Zhao Naxun

2. 经过约 28 天的孵化，小鸟才能出壳。摄影／赵纳勋
After about 28 days of incubation, chicks are hatched. Photography by Zhao Naxun

3. 幼鸟需要父母 40 天左右的关怀照顾。摄影／赵纳勋
Chicks need parental care for 40 days. Photography by Zhao Naxun

4. 摄影／张永文
Photography by Zhang Yongwen

野生生命的庇护所
CHANGQING, QINLING
A Natural Shelter to Wildlife

1. 展翅欲飞——离巢前的锻炼。摄影／赵纳勋
Fluttering their wings in preparation for the first flight. Photography by Zhao Naxun

2. 摄影／赵纳勋
Photography by Zhao Naxun

3. 清澈的河流是嬉戏理羽的好场所。摄影／赵纳勋
Crystal clear rivers are good playgrounds for grooming and frolicking. Photography by Zhao Naxun

4. 森林湿地是它的家园。摄影／赵纳勋
Forested wetlands are its home. Photography by Zhao Naxun

5. 河流、稻田等湿地是主要觅食地。摄影／赵纳勋
Rivers, rice fields and other wetlands are their major feeding grounds. Photography by Zhao Naxun

野生生命的庇护所
CHANGQING, QINLING
A Natural Shelter to Wildlife

1. 泥鳅、黄鳝是它们的主要食物。摄影/赵纳勋
Loaches and eels are their major food sources. Photography by Zhao Naxun

2. 小鱼、小虾也很可口。摄影/赵纳勋
Fish and shrimps are also delicious. Photography by Zhao Naxun

3. 软体动物及水生昆虫也是美味佳肴。摄影/赵纳勋
Molluscs and aquatic insects are delicacies. Photography by Zhao Naxun

4. 归巢时结伴而行。摄影/赵纳勋
Returning to nests in fleets. Photography by Zhao Naxun

5. 繁殖期体色为灰色,又称婚羽。摄影/赵纳勋
Feathers turn gray, known as wedding feather, during the breeding season. Photography by Zhao Naxun

| 1 | 4 |
| 2 3 | 5 |

1. 与白鹭是亲密伙伴。摄影／赵纳勋
The egret is a close friend to the crested ibis. Photography by Zhao Naxun

2. 有聚群活动的习性。摄影／赵纳勋
Crested ibises are gregarious. Photography by Zhao Naxun

| 1 | 2 |

豹 Leopard *Panthera pardus*
国家一级保护动物，别名金钱豹，本区体型最大的食肉动物。野外监测中拍摄于区内柏杨坪海拔 2528 米的山脊。红外相机摄影
First-class state protected animal, also known as gold coin leopard, the largest carnivore in the nature reserve. The picture was taken at 2,528 meters above sea level when monitoring and patrolling. Infrared Photography

野生生命的庇护所
CHANGQING, QINLING
A Natural Shelter to Wildlife

四
长青的动物大家庭

Changqing, Home to the Animals

兽类
Mammals

　　长青自然保护区是秦岭野生动物资源十分富集的地区，兽类分布的突出特点是种群数量大、种类多。有国家一级重点保护动物 5 种，二级重点保护动物 10 种。中国的特有种 12 种。大熊猫、金丝猴、羚牛、豹等标志性旗舰物种，已被广泛关注和重视，小型兽类的丰富性以及广泛分布也给长青保护区生态系统注入勃勃生机。

　　The Changqing Nature Reserve is an area with Qinling's richest wildlife resources, marked by greater and more diverse mammal populations. Among them are 5 Class I state protected species, 10 Class II state protected species and 12 species native to China. Iconic flagship species such as the giant panda, the golden monkey, the takin and the leopard gather great attention while the diversity and extensive distribution of small mammal species add to the vibrancy of the Changqing Nature Reserve's local ecosystem.

野生生命的庇护所
CHANGQING, QINLING
A Natural Shelter to Wildlife

1. 小麂（雌）Reeves' Muntjac (female) *Muntiacus reevesi*
小型的鹿科动物，多活动于中低山稀疏林地的灌丛草地。摄影／向定乾
A small deer, active in the shurb-lands of submontane sparse trees and shrublands. Photography by XiangDingqian

2. 红白鼯鼠 Red and White Giant Flying Squirrel *Petaurista alborufus*
红白色，形似松鼠，繁殖于树洞，栖息于海拔1000米左右的高大乔木林中，可由高向低滑翔，又称为飞鼠。摄影／董伟
Red and white giant flying squirrel breeds in caves of tree trunks and dwells in woody tree forests about 1,000 meters above sea level. It is also known as the flying squirrel. Photography by Dong Wei

3. 黄喉貂 Yellow-Throated Marten *Martes flavigula*
又叫青鼬，国家二级重点保护动物，穴居于树洞和岩洞，从区内海拔1000米左右的阔叶林带到近3000米的针叶林均能见到其活动的踪迹。红外相机摄影
Known also as kharza, the yellow-throated marten is a Class II state protected species that lives in trees and rock caves. It is active in needleleaf forests at 3,000 meters above seal level as well as in broad-leaved forests 1,000 meters above seal level. Infrared Photography

4. 小麂（雄）Reeves' Muntjac (male) *Muntiacus reevesi*
摄影／向定乾
Photography by Xiang Dingqian

5. 林麝 Chinese Forest Musk Deer *Moschus berezovskii*
国家一级重点保护动物，以分泌麝香而闻名，体长70厘米左右，主要栖息于中低山林区。红外相机摄影
Class I state protected species. Famous for the musk scents it releases, Chinese forest musk deer measures about 70 centimeters long and dwells in submontane forests. Infrared Photography

| 1 | 2 | 5 |
| 3 | 4 | |

野生生命的庇护所
CHANGQING, QINLING
A Natural Shelter to Wildlife

1. 中华竹鼠 Chinese Bamboo Rat *Rhizomys sinensis*
别名竹鼠，穴居，昼伏夜出，生活在海拔1000～2600米的阔叶或针阔叶混交林区内，以竹子及其地下茎为食。摄影／赵纳勋
Also known as bamboo rat, this mammal is cavernicolous and nocturnal animals. It lives 1,000 to 2,600 meters above sea level, in needleleaf forests or mixed needleleaf-broadleaf forests. It feeds on bamboos or plant roots. Photography by Zhao Naxun

2. 灰头小鼯鼠 Grey-headed Flying Squirrel *Petaurista caniceps*
昼伏夜出，喜晨昏活动，保护区内栖居于海拔1000～1500米的针阔混交林。摄影／向定乾
Crepuscular, the grey-headed squirrel is mostly active at dawn or dusk, and lives within the mixed broadleaf-needleleaf forests at 1,000 to 1,500 meters in altitude. Photography by Xiang Dingqian

3. 隐纹花松鼠 Swinhoe's Striped Squirrel *Tamiops swinhoei*
栖息于海拔较高的针叶林、林缘和灌丛，繁殖于树洞，主要以各种果实和昆虫为食。摄影／赵纳勋
Swinhoe's striped squirrel inhabits the higher altitude needleleaf forests, near timberlines and in shrublands. It reproduces in tree holes and feed on fruits, nuts and insects. Photography by Zhao Naxun

4. 斑羚 Chinese Goral *Naemorhedus griseus*
国家二级重点保护动物，俗称青羊，中等体型，区内地势陡峭的针叶林和针阔混交林常见，以嫩枝叶和青草为食。摄影／赵纳勋
Locally known as the green goat, the Chinese goral is a Class II state protected species. It is medium in build and lives within the nature reserve on the steep slopes of needleleaf forests and mixed broadleaf-needleleaf forests. It feeds primarily on fresh leaves and grass. Photography by Zhao Naxun

5. 黑熊 Asian Black Bear *Ursus thibetanus*
国家二级重点保护动物，俗称狗熊、熊瞎子，杂食性动物，有冬眠的习性，区内分布各种生境。红外相机摄影
Class II state protected animal species. Commonly known as moon bear, the Asian black bear is an omnivore who hibernates in winter and active in various habitats in the nature reserve. Infrared Photography

秦岭长青 野生生命的庇护所
CHANGQING, QINLING
A Natural Shelter to Wildlife

1. 林猬 Hugh's Hedgehog *Mesechinus hughi* Thomas
体型比手掌略大，全身长有硬刺，遇到危险时会卷成一团，性情温顺，触觉与嗅觉发达，喜食蚂蚁。摄影／胡万新
Slightly larger than the human palm, the back of a hedgehog is entirely covered with stiff spines. When attacked or in fright, it rolls into a tight ball pointing its spines out for defense. It is docile and has a sharp sense of touch and hearing. Ants are its favorite food. Photography by Hu Wanxin

2. 藏鼠兔 Moupin Pika *Ochotona thibetana*
体小，耳短圆，常见于区内中高海拔针叶林下灌丛，穴居。摄影／赵纳勋
Small in size with short rounded ears, the Moupin pika is found in the nature reserve within medium high altitudes, in the shrublands of needleleaf forests. It is a burrow dweller. Photography by Zhao Naxun

3. 鼬獾 Chinese Ferret - Badger *Melogale moschata*
头大，猪鼻，在受到威胁时会释放臭气，栖息于区内中低山林区，昼伏夜出。摄影／胡万新
With a large head and a snout, the Chinese ferret-badger releases a foul odor when threatened. It inhabits within submontane forests of the nature reserve and is a nocturnal mammal. Photography by Hu Wanxin

4. 野猪 Wild Boar *Sus scrofa*
又称山猪，区内外广泛分布，大多集群活动，食性较杂。红外相机摄影
Also known as wild pig, the wild boar is found throughout the nature reserve. It lives in large groups feeding on both plants and animals. Infrared Photography

5. 花鼠 Siberian Chipmunk *Tamias sibiricus*
行动敏捷，善于将采集的果实类食物贮存起来，但由于记忆力不强，反而对植物起到了"播种"作用。摄影／向定乾
The Siberian chipmunk is agile and keen at gathering fruits and nuts before storing them up. However, because its short memory spans, its efforts often helps disperse seeds instead. Photography by Xiang Dingqian

6. 毛冠鹿 Tufted Deer *Elaphodus cephalophus*
额顶部有一簇马蹄状黑色长毛，是以称为毛冠鹿，角短小。针阔混交林、阔叶林和疏林地、灌丛是其活动场所。摄影／赵纳勋
The name of the tufted deer is derived from the unique tuft of horseshoe-shaped hair on its forehead with small antlers hidden in them. Tufted deer is active in mixed broadleaf-needleleaf forests, broadleaf forests, sparse trees and shrublands. Photography by Zhao Naxun

7. 苏门羚 Chinese Serow *Capricornis milneedwardsii*
国家二级重点保护动物，又叫鬣羚、"四不像"，区内广泛分布，草食性动物。红外相机摄影
Class II state protected species. Mainland serow is a goat antelope which is locally nicknamed "sibuxiang," meaning "none of the four" (horse, antelope, camel and asinus). It is a herbivore and can be found throughout the limits of the nature reserve. Infrared Photography

8. 豪猪 Malayan Porcupine *Hystrix brachyuran*
活动于多种生境，穴居，夜行性，以植物根、茎为食。身体密布的长刺是防御的利器。红外相机摄影
Active in many different environments, this porcupine lives in underground burrows and becomes active at night. Plant roots and stalks are its primary food source. The quills covering its body are sharp weapons for self-defense. Infrared Photography

1. 黄鼬 Siberian Weasel *Mustela sibirica*

俗名黄鼠狼，体内具有臭腺，可排出臭气，以抵御天敌，区内见于中低山的林缘、河谷及灌丛中。摄影／赵纳勋

Commonly known as the kolonok, the Siberian weasel has odor glands in its body to release a foul odor for defense against predators. It is often found in the lower timberlines, ravines and shrublands of the nature reserve. Photography by Zhao Naxun

2. 金猫 Asian Golden Cat *Pardofelis temminckii*

国家二级重点保护动物，中等体型的食肉动物，体健有力，行踪诡秘，昼伏夜出，极难见到。红外相机摄影

Class Ⅱ state protected species. This wild cat is a medium-sized carnivore with a strong physique. Due to its crepuscular and nocturnal tendencies, it is not easily detected. Infrared Photography

野生生命的庇护所
CHANGQING, QINLING
A Natural Shelter to Wildlife

3. 猪獾 Hog-badger *Arctonyx collaris*
区内广泛分布，穴居，叫声似猪，视觉差，嗅觉发达。红外相机摄影
Found throughout the nature reserve, the hog-badger lives in burrows and snorts like pigs do. With weak vision and hearing, it has a sharper sense of smell. Infrared Photography

4. 珀氏长吻松鼠 Perny's Long-nosed Squirrel *Dremomys pernyi*
常见于区内中低山的阔叶或针阔混交林，多在树上活动。摄影／向定乾
Living in the nature reserve's mid-low broadleaf or mixed broadleaf-needleleaf forests, this squirrel is found mostly in trees. Photography by Xiang Dingqian

5. 草兔 Tolai Hare *Lepus tolai*
低海拔林地和社区周围广泛分布。摄影／胡万新
Cape hare is widely distributed in low-altitude woodlands and surrounding local communities. Photography by Hu Wanxin

6. 花面狸 Masked Palm Civet *Paguma larvata*
别名果子狸，果实丰富的森林是它的家，家族式活动，昼伏夜出，食性较杂。红外相机摄影
Sometimes known as gem-faced civet, this small mammal settles its home in forests rich with food sources such as fruits and nuts. It is a nocturnal omnivore. Infrared Photography

7. 豹猫 Leopard Cat *Prionailurus bengalensis*
体型与家猫相仿，保护区内分布广泛，肉食性。摄影／赵纳勋
With similar physical characteristics of a house cat, the leopard cat is a carnivore that can be found throughout the nature reserve grounds. Photography by Zhao Naxun

巫山角蟾 Wushan Horned Toad *Megophrys wushanensis*
腹部和四肢腹面为酱色，生活于海拔 1000 米左右的溪流及其附近，6 月中旬开始产卵孵化。摄影／向定乾
The Wushan horned toad has dark brown abdomen and limbs. It lives in streams around 1,000 meters in altitude. It reproduces in mid June. Photography by Xiang Dingqian

两栖爬行动物
Amphibians & Reptiles

处于北亚热带与暖温带分界线上的长青保护区，沟谷纵横，河流众多。气候季节性变化明显，具有雨热同季、温暖湿润、雨量充沛等特点，并随海拔的升高呈垂直变化。加之植被类型和生境类型多样，小气候差异明显，使长青保护区成为两栖、爬行动物生存的理想之地。目前，已发现两栖动物 13 种，爬行动物 22 种。

Located on the borders of the northern subtropical zone and the temperate zone is the Changqing Nature Reserve, an ecological area criss-crossing with valleys, ravines and rivers. The seasons are distinct and the climate is mild, warm, and moist with ample rainfall, all due to the altitudinal variation of the area. Additionally, the diverse vegetation, various types of habitats, along with prominent changes in the microclimate, make Changqing an ideal home to amphibians and reptiles. Thus far, 13 amphibian species and 22 reptile species have been discovered in the nature reserve.

秦岭长青 野生生命的庇护所
CHANGQING, QINLING
A Natural Shelter to Wildlife

1	3	8
2	4	
5	6	7

1. 北草蜥 Northern grass lizard *Takydromus septentrionalis*
生活在中低海拔山地草丛中，体细长，尾为体长的两倍以上。摄影／赵纳勋
Living in the mid-low altitude mountain and in the grass field is this slim-bodied lizard whose tail measures twice its body length. Photography by Zhao Naxun

2. 北方山溪鲵 Western Chinese mountain salamander *Batrachuperis pinchonii*
生活在中高海拔溪流中，以藻类和水生昆虫为食。摄影／赵纳勋
Living in the streams of mid-high altitudes, the western Chinese mountain salamander feeds on algae and aquatic insects. Photography by Zhao Naxun

3. 秦岭滑蜥 Tsinling Dwarf Skink *Scincella tsinlingensis*
生活在区内中高海拔的向阳山坡，多活动于灌丛或草丛间，亦隐匿于乱石下，主要捕食小型昆虫。摄影／李成
Living on the nature reserve's mid- to high-altitude sun-facing slopes, Tsinling dwarf skink is active in shrubs, grass fields and under rocks, mostly feeding on small insects. Photography by Li Cheng

4. 大鲵 Chinese Giant Salamander *Andrias davidianus*
世界上现存最大的两栖动物，国家二级重点保护动物，中国特有种。因叫声似幼儿哭声，故俗称"娃娃鱼"，栖息于区内中低海拔河流中。摄影／向定乾
As the largest living amphibian in the world, the Chinese giant salamander is a Class II state protected species and indigenous to China. Since the salamander vocalizes and produces sounds much like those of a baby, it is given a local name: infant fish. It mostly inhabits lower-altitude rivers within the nature reserve. Photography by Xiang Dingqian

5. 米仓山攀蜥 Micang Mountain Japalure *Japalura micangshanensis*
见于海拔800米左右的山坡草丛。摄影／董伟
The Micang mountain japalure is seen on grass slopes 800 meters above sea level. Photography by Dong Wei

6. 丽纹攀蜥 Splendidjapalure Splendid Japalure *Japalura splendida*
日行性攀蜥，栖息于海拔较高的山区，以各种小型昆虫及其幼虫为食。摄影／李成
This diurnal lizard inhabits the higher-altitude mountains, eating small insects and their larvae. Photography by Li Cheng

7. 蓝尾石龙子 Five-Striped Blue-Tailed Skink *Eumeces elegans*
栖息于区内低海拔地区，喜欢在干燥而温度较高的阳坡活动，以昆虫为主食。摄影／李成
Five-striped blue-tailed skink lives on the dry, warm, sun-facing mountain slopes at the lower altitudes of the nature reserve. Insects are its main source of food. Photography by Li Cheng

8. 铜蜓蜥 Brown Forest Skink *Sphenomorphus incognitus*
栖息于海拔2000米以下的中低海拔地区。体背古铜色并有一条黑脊纹，体两侧各有一黑色纵带。摄影／赵纳勋
Brown forest skink can be found in the mid-low altitude areas under 2,000 meters. It is identified by its tan-colored back marked by a black line along the center and two additional stripes on the sides. Photography by Zhao Naxun

1. 虎斑颈槽蛇 Tiger Keelback *Rhabdophis tigrinus*
微毒，体长近1米，区内分布于中低海拔地区，受惊扰或被激怒时能昂首举颈，颈部膨扁，身体呈"S"形弯曲。摄影／赵纳勋
Slightly venomous, the tiger keelback is one-meter long and is distributed in mid-low altitudes. Once startled or angered, it raises its head, raising and flattening its neck, and archs itself in place for attack. Photography by Zhao Naxun

2. 翠青蛇 Greater Green Snake *Cyclophiops major*
非常温顺的无毒蛇，喜攀爬上树，静伏纳凉，体细长，约1米左右。摄影／赵纳勋
A docile nonvenomous snake, the greater green snake is a keen tree climber that chooses to stay and cool off in the shades. It has long thin bodies roughly a meter in length. Photography by Zhao Naxun

3. 颈槽蛇 Keelback Snake *Rhabdophis*
生活在海拔1500~1800米的林区，以蚯蚓、蛞蝓等为食，头较大，与颈区分明显。摄影／胡万新
Inhabiting the forests 1,500 to 1,800 meters above sea level, and feeding on insects such as earthworms and slugs, the newborn keelback snake has a large heads relative to its neck. Photography by Hu Wanxin

4. 黑脊蛇 Black-spine Spinal Snake *Achalinns spinalis*
无毒蛇，体背面棕黑色，背中央有一条醒目的黑脊线，主食蚯蚓。摄影／胡万新
Black-spine spinal snake is a nonvenomous species with black brown scales. Along the middle of its back runs a clearly marked, black line. The main food source for the snake is the earthworm. Photography by Hu Wanxin

5. 黑眉锦蛇 Beauty Snake *Elaphe taeniura*
区内常见无毒蛇，眼后有一条明显的黑纹，经常在小动物出没的地方游动、捕食。摄影／赵纳勋
The beauty snake is a commonly-seen nonvenomous snake in the nature reserve. It has distinct black marks near its eyes and makes its appearance where small animals are active. Photography by Zhao Naxun

6. 菜花烙铁头 Jerdon's pitviper *Trimeresurus jerdonii*
秦岭剧毒蛇，背面草绿色，杂以黄红及黑色斑点，中高海拔区域常见。摄影／赵纳勋
A highly venomous snake native to Qinling, the Jerdon's pitviper has a green back line with orange and black patterns. This snake appears most often in mid-high altitudes. Photography by Zhao Naxun

7. 紫灰锦蛇 Red Bamboo Ratsnake *Elaphe porphyracea*
无毒蛇，体长约1米左右，生活于中低海拔林缘、田地、河畔及居民点，以小型哺乳类为食。摄影／赵纳勋
Red bamboo ratsnake is nonvenomous snakes measuring one meter in length and active in the mid-low altitude timberlines, fields, by the rivers or near communities. It preys on small mammals for food. Photography by Zhao Naxun

野生生命的庇护所
CHANGQING, QINLING
A Natural Shelter to Wildlife

秦岭长青 野生生命的庇护所
CHANGQING, QINLING
A Natural Shelter to Wildlife

1. 秦岭蝮 Qinling Pit-viper *Gloydius qinlingensis*
秦岭的剧毒蛇之一，当地居民又称其为土蝮子。多见于区内中低海拔林间空地、岩石和灌草丛。摄影／赵纳勋
One of the most venomous snakes in Qinling, this snake is also known by natives as dirt viper. It is notably found in the nature reserve's mid-low altitudes, in empty fields, rock crevices and grass fields. Photography by Zhao Naxun

2. 斜鳞蛇 Oblique-scale snake *Pseudoxenodon macrops*
身具臭味，故称臭蛇，无毒，受惊时身体前段竖起，颈膨扁，中低海拔区常见。摄影／赵纳勋
Locally known as the stinking snake for its unpleasant body odor, the obliquescale snake will raise the front section of its body when startled while expanding and flattening its neck. It is found at the mid-low altitudes of the nature reserve. Photography by Zhao Naxun

3. 乌梢蛇 Big Eye Snake *Zoacys dhumnades*
体型较大的无毒蛇，保护区内中低海拔区域广泛分布，以鱼、蜥蜴、蛙类及鼠类为食。摄影／向定乾
A slightly larger nonvenomous snake, the big eye snake has widespread populations in the nature reserve. It preys on fish, lizards, frogs and small rodents. Photography by Xiang Dingqian

4. 隆肛蛙 Swelled Vent Frog *Nanorana quadranus*
栖息于保护区内1800米以下的溪流、水坑及近水的草地、灌丛。摄影／胡万新
Swelled vent frog lives in brooks, pools, grasses or shrubs close to the water, mostly under 1,800 meters in altitude. Photography by Hu Wanxin

5. 华西蟾蜍 West China Toad *Bufo andrewsi*
体表色似枯树叶、泥土，具有毒的腺体，栖居于草丛、石下或土洞中，3～5月产卵，以昆虫为食。摄影／胡万新
Camouflaged with colors similar to dead leaves and mud, the west China toad has venomous glands. It dwells in grass bushes, under rocks or in soil caves. It lays eggs from March to May and feed on insects. Photography by Hu Wanxin

6. 中华蟾蜍 Zhoushan Toad *Bufo gargarizans*
俗名癞疙疱、癞蛤蟆。冬眠和繁殖期在水中生活，其余时间活动于林下、草丛。摄影／向定乾
Locally known as scabby toad, the Zhoushan toad lives in the water while hibernating or during reproduction. During other times, it lives under tree shades or in the grass fields. Photography by Xiang Dingqian

1. 合征姬蛙 Mixtured Pygmy Frog *Microhyla mixtura*
中国的特有物种，体小2厘米左右。生活于中低海拔稻田、水坑及附近，捕食昆虫等动物性食物。摄影／向定乾
The mixtured pygmy frog is an amphibian species endemic to China. Its minuscule body measures only 2 centimeters long. The frog dwells in mid-low altitude paddy fields, pools and nearby, preying on small insects for food. Photography by Xiang Dingqian

2. 秦岭雨蛙 Tsinling Tree Toad *Hyla tsinlingensis*
中国特有物种，又称秦岭树蟾，栖息于中低海拔稻田或水域附近。体型娇小，能吸附在树叶、草叶上，捕食小昆虫。摄影／向定乾
Endemic to China, the Tsinling tree toad inhabits the mid-low altitude paddy fields or nearby bodies of water. It is small and capable of suctioning onto tree leaves and grass. Small insects are its favorite food. Photography by Xiang Dingqian

3. 中国林蛙 Chinese Forest Frog *Rana chensinensis*
头体和四肢较细长，行动敏捷，跳跃力强。食物主要为鞘翅类昆虫。摄影／向定乾
With a slender body and limbs, this frog is agile and a powerful jumper. It feeds mainly on beetles. Photography by Xiang Dingqian

4. 黑斑侧褶蛙 Black-spotted Frog *Pelophylax nigromaculatus*
俗称"青蛙"、"田鸡"。常栖息于稻田、池塘、水沟或近水的草丛中。摄影／向定乾
Locally known as green frogs, the black- spotted frog dwells in paddy fields, ponds, ditches or rice fields with ready access to water. Photography by Xiang Dingqian

秦岭细鳞鲑 Qinling Lenok *Brachymystax lenok tsinlingensis*
国家二级保护动物，中国特有物种。生活在海拔 900～2300 米的山涧河流中，肉食性冷水鱼，早晚及阴天捕食频繁。摄影／向定乾
Class II state protected species. Endemic to China, the Qinling lenok lives 900 to 2,300 meters above sea level in the ravines and rivers. It is a carnivorous coldwater fish which feeds at dawn, dusk and on cloudy days. Photography by Xiang Dingqian

鱼类
Fish

长青保护区地处汉江流域，也是汉江一级支流酉水河的发源地，汇集了区内 7 条支流，流域面积 280 多平方千米；胥水河的支流大西河也发源于境内。丰富的淡水资源孕育了多样的鱼类，鱼类区系组成的主要成分属于长江中游鱼类区系，有 5 目 18 种。

The Changqing Nature Reserve is located along the banks of the Han River and is the source of its main river branch Youshui River. The nature reserve encompasses seven river branches, covering 280 square kilometers of river surface. Meanwhile, the headwaters of the Daxi River which diverges from the Xushui River, also originate from here. The plentiful fresh water resources nurtures abundant fish fauna which contributes to the regional fauna of the middle reaches of the Yangtze River. These contain 5 fish orders and 18 species.

野生生命的庇护所
CHANGQING, QINLING
A Natural Shelter to Wildlife

1. 似鮈 *Pseudogobio vaillanti*
小型鱼类，生活于河流下层。体长而背部隆起，腹部平坦。摄影／胡万新
This small fish lives at the bottom of river beds. It has a long figure, a thicker torso and a flat abdomen. Photography by Hu Wanxin

2. 马口鱼 Chinese Hooksnout Carp *Opsariichthys bidens*
俗名桃花鱼，体型小，常见于水流较急的浅滩和砂砾多的小河中。性较凶猛，肉食性鱼类。摄影／赵纳勋
Locally known as peach blossom fish, the Chinese hook snout carp is a small fish found in riffles or pebbled rivers. It is an aggressive carnivore. Photography by Zhao Naxun

3. 鲫 Crucian Carp *Carassius auratus auratus*
见于区内及附近较大河流和池塘中。草食性为主，冬季多潜入水底深处越冬。摄影／胡万新
Found in the nature reserve in larger rivers and ponds, the crucian carp is a herbivore and hibernates at the bottom of the water during winter. Photography by Hu Wanxin

4. 宽鳍鱲 Pale chub *Zacco platypus*
体小，见于区内较大溪流中。以浮游甲壳类为食，兼食一些藻类、小鱼及水底的腐殖物质。摄影／胡万新
A small fish found in larger streams within the nature reserve, the pale chub eats small crustaceans as its primary diet along with macroscopic algae, small fish and detritus. Photography by Hu Wanxin

5. 红尾副鳅 *Paracobitis variegatus*
体细长、圆柱状，喜栖息在岩缝、石隙或多石的洄水湾，常以下颌发达的角质边缘在岩石上刮取食物。摄影／胡万新
Slender and cylindrical in shape, this fish hides in the crevices of rocks, stones and in the microbe-rich waters of pebbled bays. It uses the cuticle of its lower jaw to scrape organic matters from rock surfaces for food. Photography by Hu Wanxin

6. 拉氏鲅 *Phoxinus lagowskii*
冷水性鱼类，集群活动于保护区的河流、小溪。摄影／赵纳勋
Cold water fish, the shoal is commonly seen in the river and creek in the nature reserve. Photography by Zhao Naxun

7. 花䱻 Spotted Steed Fish *Hemibarbus maculatus*
生长较慢，广泛分布于区内及周边较大河流溪谷中，喜欢在水的中下层活动。摄影／胡万新
Spotted steed fish have a slow growth rate and are widely found in local rivers and ravines. They are active near the bottom of the river bed. Photography by Hu Wanxin

红腹角雉 Temminck's Tragopan *Tragopan temminckii*
红外相机摄影 Infrared Photography

野生生命的庇护所
CHANGQING, QINLING
A Natural Shelter to Wildlife

鸟类
Birds

　　繁茂的植被，多样的生境，宜人的气候、较大的落差以及独特的地理位置，使长青保护区成为鸟类栖息的乐园。现已观察记录到鸟类319种，其中国家一级保护鸟类4种，二级保护鸟类40种。鸟类中有中国特有鸟16种。留鸟是保护区鸟类区系的主要构成成分。种类繁多的鸟类在维持长青保护区生态系统稳定性方面发挥着重要作用，也是从事鸟类研究和观鸟的理想之地。

　　Rich vegetation, great biodiversity, mild climate, distinct seasonal changes and a unique geographical location render the Changqing Nature Reserve an ideal habitat grounds to birds. 319 bird species have been observed so far and among them are 4 Class I state protected species and 40 Class II state protected species. Among these, 16 species are found to be endemic to China, and non-migratory birds make up the majority of the reserve ground's avifauna. The variety of birds plays an important role in conserving Changqing's biodiversity, making it an ideal place for studying birds and for birdwatching.

陆禽

Ground Birds

陆禽常在陆地觅食植物的种子、叶芽和根茎，也取食昆虫和其他小型动物。长青保护区有陆禽15种，大多为留鸟。其中鸡形目中的血雉、红腹角雉、红腹锦鸡、勺鸡和灰胸竹鸡在区内分布广、种群大、数量多，为长青保护区优势种。陆禽中有国家二级保护动物6种。

Ground birds forage for food on the forest grounds, pecking at plant seeds, tender shoots and plant roots and some also feed on insects and small animals. The Changqing Nature Reserve is home to 15 species of ground birds, most of which are resident birds. Among them, galliformes such as blood pheasants, Temminck's tragopans, golden pheasants, the Koklass pheasants and Chinese bamboo partridges are considered Changqing's dominant species, as the largest in population and widest in distribution. The ground bird populations include 6 Class Ⅱ state protected bird species.

摄影 / 赵纳勋 Photography by Zhao Naxun

1. 环颈雉（雄）Common Pheasant (male) *Phasianus colchicus*
又名野鸡，山鸡，雉鸡，雄性色泽艳丽。中低海拔的浅山、灌丛及农田周边广泛分布，四季常见。摄影／赵纳勋
Locally known as the mountain chicken, the common pheasant are widely distributed in mid-low altitude hills, shrublands and surrounding fields. The male bird has vibrant showy colors. It can be seen throughout the year. Photography by Zhao Naxun

2. 珠颈斑鸠 Spotted Dove *Spilopelia chinensis*
留鸟。颈侧黑色斑块中的白点似珍珠。区内外中低海拔河谷田地广泛分布。摄影／赵纳勋
This dove is a buff brown non-migratory bird with a pearl-spotted black collar patch on the back and sides of the neck. In the nature reserve, they are widely distributed in the mid-low altitude ravines and fields. Photography by Zhao Naxun

3. 山斑鸠 Oriental Turtle Dove *Streptopelia orientalis*
留鸟，为区内常见种。与珠颈斑鸠的区别在于颈侧有带明显的黑白色条纹的块状斑。摄影／胡万新
This dove differs from the spotted dove in that it has striated black and white stripes on the side of its necks, instead of spots. The oriental turtle dove is a frequently sighted resident species in the nature reserve. Photography by Hu Wanxin

4. 环颈雉（雌）Common Pheasant (female) *Phasianus colchicus*
摄影／向定乾 Photography by Xiang Dingqian

1. 红腹角雉（雌）Temminck's Tragopan (female) *Tragopan temminckii*
世界濒危鸟类，国家二级保护动物。保护区海拔1400～2600米处常见。色彩绚丽而富于变幻，叫声奇特。红外相机摄影
Class Ⅱ state protected bird species. Male Temminck's tragopans are often seen in the nature reserve's at an altitude of 1,400 to 2,600 meters. They show striking colors on the plumage and have a unique call. Infrared Photography

2. 红腹角雉（雄）Temminck's Tragopan (male) *Tragopan temminckii*
摄影／向定乾 Photography by Xiang Dingqian

3. 血雉（雄）Blood Pheasant (male) *Ithaginis cruentus*
国家二级保护动物，又称松花鸡，本区优势种，种群大，数量多。栖息于海拔2000米以上的针叶林、混交林和杜鹃灌丛间。摄影／向定乾
Class Ⅱ state protected bird species. The blood pheasant is a dominant species in the area, both in population and in number. They inhabit 2,000 meters above sea level, in high needleleaf forests, mixed forests and rhododendron brushes. Photography by Xiang Dingqian

4. 血雉（雌）Blood Pheasant (female) *Ithaginis cruentus*
摄影／向定乾 Photography by Xiang Dingqian

5. 火斑鸠 Red Collared Dove *Streptopelia tranquebarica*
本地为夏候鸟。体色呈酒红色，见于中低海拔区。摄影／赵纳勋
A summer migrant bird in Changqing, the red collared dove has wine-red plumage and can be seen at mid-low altitudes within the nature reserve. Photography by Zhao Naxun

6. 灰斑鸠 Eurasian Collared Dove *Streptopelia decaocto*
留鸟。褐灰色，特征为后颈有黑白色半领圈，见于中低海拔河谷山地。摄影／赵纳勋
A non-migratory bird in the local area, the Eurasian collared dove is grey-buff in color, with a black half-collar edged with white on its nape. They are mostly seen at mid-low altitude river valleys and mountains. Photography by Zhao Naxun

秦岭长青 野生生命的庇护所
CHANGQING, QINLING
A Natural Shelter to Wildlife

1. 红腹锦鸡 Golden Pheasant *Chrysolophus pictus*
国家二级保护动物，别名金鸡。羽色华丽，为中国特有种，中低海拔广泛分布。摄影／赵纳勋
Class II state protected species. Locally known as the golden chicken, this ground bird has gorgeous colors, and is endemic to China and widely distributed at the mid-low altitudes of the reserve. Photography by Zhao Naxun

2. 灰胸竹鸡 Chinese Bamboo Partridge *Bambusicola thoracicus*
中国特有种。在本区数量多、分布广，中低海拔区常见。摄影／赵纳勋
Endemic to China, the Chinese bamboo partridge has a large population and is widely distributed at the mid-low altitudes. Photography by Zhao Naxun

3. 白冠长尾雉 Reeves's Pheasant *Syrmaticus reevesii*
留鸟。国家二级保护动物，中国特有种。雄鸟具斑斓的体色，加上尾羽可达1.7米左右。数量少，偶见于低海拔地区山王庙等地。摄影／华英
Class II state protected species. This bird is a resident species endemic to China. The male bird's colorful body and tail can measure up to 1.7 meters long. They are few in number and can occasionally be seen at lower altitudes such as Shanwangmiao area. Photography by Hua Ying

4. 勺鸡 Yellow-necked Koklass Pheasant *Pucrasia macrolopha*
国家二级保护动物，长青保护区常见种，数量较多。摄影／向定乾
Class II state protected species. These birds are commonly seen in the Changqing Nature Reserve. Photography by Xiang Dingqian

5. 点斑林鸽 Speckled Wood Pingon *Columba hodgsonii*
留鸟。与其他鸽种的区别在于颈部羽毛形长而具端环，栖息于区内中高海拔多悬崖峭壁的森林。摄影／赵纳勋
Non-migratory, this pigeon differs from others in that the plumage on its chest extends to the back of its neck like a band. These birds inhabit the rugged cliffs of mid-high altitude forests. Photography by Zhao Naxun

1		4	5
2	3		

黄脚渔鸮 Tawny Fish Owl *Bubo flavipes*
摄影／赵纳勋 Photography by Zhao Naxun

猛禽

Birds of Prey

猛禽为性格凶悍的肉食性鸟类，以捕食鸟、兽为食。包括隼形目和鸮形目的所有种。长青保护区目前已发现猛禽33种，其中国家一级保护2种，二级保护31种。

Birds of prey are aggressive carnivorous hunters, preying on other birds and animals for food. They include all the species in the orders Falconiformes and Strigiformes. By 2012, 33 species of birds of prey have been sighted in the Changqing Nature Reserve, including 2 Class Ⅰ state protected species and 31 Class Ⅱ state protected species.

1	2
3	4

1. 鹰雕 Mountain Hawk-Eagle *Nisaetus nipalense*
国家二级保护动物。常在阔叶林和混交林中活动，冬季在海拔较低的阔叶林和林缘地带活动。红外相机摄影
Class II state protected species. These birds are mostly active in broadleaf forests and mixed forests, but prefer to winter in lower altitude broadleaf forests and near timberlines. Infrared Photography

2. 斑头鸺鹠 Barred Owlet *Glaucidium cuculoides*
留鸟。国家二级保护动物。体型较小，常见于保护区中低海拔林区及社区林缘。摄影／赵纳勋
Class II state protected species. This non-migratory owl is small in size and found in the mid-low altitude forests and timberlines surrounding local communities. Photography by Zhao Naxun

3. 赤腹鹰 Chinese Sparrowhawk *Accipiter soloensis*
候鸟。小型猛禽，国家二级保护动物。于本地繁殖，春夏季常见于保护区针阔混交林及周边社区林缘。摄影／赵纳勋
Class II state protected species. Chinese sparrowhawks are small migrant predators. They breed locally and can be found in the spring and summer, in mixed broadleaf-needleleaf forests and timberlines surrounding built communities. Photography by Zhao Naxun

4. 大鵟 Upland Buzzard *Buteo hemilasius*
冬候鸟，国家二级保护动物。体型较大(70厘米)的棕色鵟，尾上偏白并常具横斑。区内及周边偶见。摄影／向定乾
A winter migrant and a Class II state protected species, the upland buzzard is a larger (70 centimeters) brown bird with a paler tail often marked with striated patterns. They can often be seen in the nature reserve and nearby areas. Photography by Xiang Dingqian

野生生命的庇护所
CHANGQING, QINLING
A Natural Shelter to Wildlife

1. 东方角鸮 Oriental Scops Owl *Otus sunia*
留鸟。国家二级保护动物。又叫红角鸮，体型娇小，眼大黄色。低海拔社区偶见。摄影／赵纳勋
Locally known as the red horned owl, this non-migratory bird is a Class Ⅱ state protected species. It is small in size with large yellow eyes and are occasionally sighted near built communities at lower altitudes. Photography by Zhao Naxun

2. 黑冠鹃隼 Black Baza *Aviceda leuphotes*
留鸟。国家二级保护动物。在猛禽中体型较小，头顶直立的黑色长冠羽显现。以大型昆虫为食，常见于保护区杨家沟、吊坝河一带。摄影／赵纳勋
Non-migratory, this Class Ⅱ state protected bird is relatively small with a long, prominent crest on its head. They feed mainly on insects and are found within the nature reserve near Yangjiagou and the Diaoba River. Photography by Zhao Naxun

3. 红隼 Common Kestrel *Falco tinnunculus*
留鸟。国家二级保护动物，为体型较小的赤褐色隼，见于保护区中低海拔区域。摄影／赵纳勋
The common kestrel is a non-migratory, Class Ⅱ state protected bird with a small build and chestnut-colored plumage. They can be found in the mid-low altitudes within the nature reserve. Photography by Zhao Naxun

4. 灰脸鵟鹰 Grey-faced Buzzard Eagle *Butastur indicus*
过境鸟。国家二级保护动物。为中等体型的偏褐色鵟鹰，偶见于保护区及周边中低海拔区域。摄影／赵纳勋
Class Ⅱ state protected species. This medium-sized passage migrant has brownish plumage. They are occasionally sighted in the nature reserve and its surrounding mid-low altitude regions. Photography by Zhao Naxun

5. 雕鸮 Eurasian Eagle Owl *Bubo bubo*
留鸟。国家二级保护动物。夜行性猛禽，中国体型最大的猫头鹰。眼大橙色，耳羽发达。善于夜晚在林间飞行。摄影／赵纳勋
Class Ⅱ state protected species. The Eurasian eagle owl is a nocturnal predator and the largest owl in China. It has large orange eyes and thick ear tufts. It is a non-migratory bird which is keen on taking flight at night. Photography by Zhao Naxun

6. 黄脚渔鸮 Tawny Fish Owl *Bubo flavipes*
留鸟。全球近危物种，国家二级保护动物。常见于保护区森林茂密的河流两岸，半昼行性捕鱼为食。摄影／张永文
This non-migratory bird is a Class Ⅱ state protected bird species found in river banks along the dense forests within the nature reserve. They prey on fish as its primary diet, and usually hunt in the afternoon. Photography by Zhang Yongwen

7. 凤头鹰 Crested Goshawk *Accipiter trivirgatus*
留鸟。国家二级重点保护动物。栖息于密林覆盖处，繁殖期常在森林上空翱翔。摄影／赵纳勋
The crested goshawk is a non-migratory, Class Ⅱ state protected bird species inhabiting thick foliaged forests. During the mating season they are often seen gliding above the forest. Photography by Zhao Naxun

1	3
	5
2	4

1. 金雕 Golden Eagle *Aquila chrysaetos*
留鸟。国家一级保护动物。体型较大，头具金色冠羽。性情凶猛，体态雄伟，分布于保护区中高海拔林区。摄影／赵纳勋
Class Ⅰ state protected species. The golden eagle is a large non-migratory bird with a golden crown-like crest. It is an aggressive hunter with a magnificent stature. They are distributed within mid-high altitude forests within the nature reserve. Photography by Zhao Naxun

2. 领鸺鹠 Collared Owlet *Glaucidium brodiei*
留鸟。国家二级保护动物。我国体型最小（15厘米）的鸮类，后颈领斑和黑斑是其显著特征。见于保护区中低海拔及周边林区。摄影／赵纳勋
Class Ⅱ state protected species. The collared owlet is non-migratory and the smallest owls (15 centimeters) living in China. It is identified by the dark patterns on the back of its neck and are often seen in the nature reserve within the mid-low altitudes and nearby forests. Photography by Zhao Naxun

3. 秃鹫 Cinereous Vulture *Aegypius monachus*
全球性近危物种，国家二级保护动物。为体型硕大的深褐色鹫，偶见于保护区及周边地区。以食腐烂动物为主，在本区为过境鸟。摄影／赵纳勋
Globally classified as near threatened, the cinereous vulture is a Class Ⅱ state protected bird in China. It is very large in build and has dark brown plumage. They can occasionally be seen in the nature reserve and its surroundings. Feeding on dead animals, this vulture is a passage migrant to the area. Photography by Zhao Naxun

4. 鹰鸮 Brown Hawk Owl *Ninox scutulata*
繁殖鸟。国家二级保护动物。无明显的脸盘和领翎。多于夜间和晨昏活动，行踪诡秘，以老鼠和鸟类为食，本区常见。摄影／赵纳勋
The brown hawk owl is a breeder bird and a Class Ⅱ state protected species. It lacks a distinct facial disk and neck. Active during the night or at dawn and dusk, this owl is mysterious and preys on rodents and birds as its primary diet. They can be seen throughout the nature reserve. Photography by Zhao Naxun

5. 松雀鹰 Besra *Accipiter virgatus*
留鸟。国家二级保护动物。常在林间枝上静立伺机捕食鼠类、小鸟、昆虫等动物。摄影／向定乾
Class Ⅱ state protected species. Besras are resident birds often seen perched quietly on tree branches preparing for surprise attacks on rodents, small birds and insects. Photography by Xiang Dingqian

1	2	5	6
3	4		

1. 黄爪隼 Lesser Kestrel *Falco naumanni*
国家二级保护动物，在本区内为不常见的候鸟。于悬崖峭壁营巢，主要以昆虫为食。摄影／胡万新
Class Ⅱ state protected species. Lesser kestrels are rare in the nature reserve. They build their nests in cliffs and feed on insects as their primary diet. Photography by Hu Wanxin

2. 阿穆尔隼 Amur Falcon *Falco amurensis*
国家二级保护动物，黄昏后捕捉昆虫，常与黄爪隼混群，喜立于电线上。摄影／胡万新
Class Ⅱ state protected species. Amur falcons become active after sunset, hunting for insects. They mix well with lesser kestrels and are often found perching on telephone wires. Photography by Hu Wanxin

3. 苍鹰 Northern Goshawk *Accipiter gentilis*
国家二级保护动物。上体青灰，下体白色具粉褐色横斑。候鸟，见于保护区中低山区及村庄周围。摄影／向定乾
Class Ⅱ state protected species. The upper body of the northern goshawk is blue grey and its lower body is white with pale brown patterns. This migratory bird can be seen in mid-low mountains or surrounding villages. Photography by Xiang Dingqian

4. 纵纹腹小鸮 Little Owl *Athene noctua*
留鸟。国家二级保护动物。体长约23厘米，以鼠类、昆虫和小动物为食，见于低海拔及周边丘陵区。摄影／张永文
Non-migratory, the little owl is a Class Ⅱ state protected species in China. It measures 23 centimeters in length and lives on rodents, insects and other small animals. They are found in lower altitudes and near foothills. Photography by Zhang Yongwen

5. 雀鹰 Eurasian Sparrowhawk *Accipiter nisus*
留鸟，体型中等，脸颊棕色为识别特征。为保护区常见种，分布广泛。常在林缘和开阔林区捕食。摄影／赵纳勋
The sparrowhawk is a resident bird of medium size and with a brown cheek. They are Commonly seen in the nature reserve. They usually prey at open space among woods and the edge of forests. Photography by Zhao Naxun

6. 普通鵟 Common Buzzard *Buteo buteo*
过境鸟。国家二级保护物种。体型略大的红褐色鵟。常见于林缘草地和村庄上空盘旋翱翔，以森林鼠类为食。摄影／赵纳勋
Class Ⅱ state protected species. The common buzzard is a larger passage migrant bird with a reddish brown plumage. It is often seen at timberlines, meadows, and gliding above villages preying on rodents as its primary food source. Photography by Zhao Naxun

燕 雀 Brambling *Fringilla montifringilla*
摄影／赵纳勋 Photography by Zhao Naxun

鸣禽

Songbirds

中、小型善于鸣唱的鸟类，雀形目的所有鸟类均为鸣禽。鸣禽是鸟类进化史中最为进步的类群，占据多种多样的生态环境，种类繁多，占鸟类的绝大多数。长青保护区的鸣禽多达180种，在区内广泛分布，种群大、数量多。

Songbirds are small singing birds. They include all species belonging to the order Passeriformes. Songbirds are the most evolved group of birds with territories and habitats covering diverse climates and ecological environments. They make up the majority of all bird species. In the Changqing Nature Reserve, 180 songbird species have been identified and they are widely distributed and large in population.

1. 八哥 Crested Myna *Acridotheres cristatellus*
留鸟。集群活动于华阳保护站周边社区，以昆虫、果实为食。摄影／赵纳勋
This non-migratory bird is active in groups near the Huayang ranger station and its surrounding communities. It feeds primarily on insects, fruits and nuts. Photography by Zhao Naxun

2. 暗灰鹃鵙 Black-winged Cuckooshrike *Coracina melaschistos*
候鸟。数量较少，不易见到，偶然发现并摄于保护区境内吊坝河。摄影／张永文
The black-winged cuckooshrike is migratory bird that has a very small population. It is rarely sighted in the region. Captured on camera by chance, this bird was spotted along the Diaoba River. Photography by Zhang Yongwen

3. 暗绿柳莺 Greenish Warbler *Phylloscopus trochiloides*
体型较小的柳莺，夏季栖息并繁殖于高海拔的灌丛及林地。摄影／向定乾
During the summer, this smaller warbler dwells and reproduces in high altitude shrublands and forests. Photography by Xiang Dingqian

4. 暗绿绣眼鸟 Japanese White-eye *Zosterops japonicus*
候鸟。常见于多种有林环境，活泼喧闹、行动敏捷、叫声婉转动听，于5、6月份繁殖。摄影／赵纳勋
The Japanese white-eye is migratory and found in diverse wooded environments. It is lively and vocally active bird. It gives out pleasing chirps and songs while remaining rather acrobatic. Its breeding season occurs in May and June. Photography by Zhao Naxun

秦岭长青
野生生命的庇护所
CHANGQING, QINLING
A Natural Shelter to Wildlife

1. 白顶溪鸲 White-capped Water-redstart *Chaimarrornis leucocephalus*
留鸟。头顶白色，腰腹部及尾基为栗色，保护区境内溪流河谷地带常见。
摄影／张永文
Non-migratory, this redstart has a white head, chestnut brown underparts and tail. They are often sighted near streams and ravines within the nature reserve. Photography by Zhang Yongwen

2. 白冠燕尾 White-crowned Forktail *Enicurus leschenaulti*
留鸟。中等体型的黑白色燕尾，河流溪谷经常能看到它鸣叫疾飞的身影。
摄影／赵纳勋
Non-migratory and medium in size, this bird has a black and white forked tail. Near rivers and ravines, they can often be seen flying by while giving out calls. Photography by Zhao Naxun

3. 白喉噪鹛 White-throated Laughingthrush *Garrulax albogularis*
留鸟。喉及上胸白色，常集群活动于林中，相互鸣叫前行，食昆虫和草籽。
摄影／赵纳勋
This non-migratory bird has a white throat and breast. They are often seen in the woods and active in groups, singing and chirping about. They feed on insects and grass seed. Photography by Zhao Naxun

4. 白喉噪鹛（雏鸟）White-throated Laughingthrush (nestlings) *Garrulax albogularis*
摄影／赵纳勋 Photography by Zhao Naxun

5. 白腹[姬]鹟 Blue-and-white Flycatcher *Cyanoptila cyanomelana*
候鸟。区内见于海拔1800米左右近山脊的针阔混交林，以昆虫为食。
摄影／向定乾
This migratory bird can be found in mixed broadleaf-needleleaf forests at approximately 1,800 meters in altitude. Insects are its primary food source. Photography by Xiang Dingqian

6. 白颊噪鹛 White-browed Laughingthrush *Garrulax sannio*
留鸟。低海拔次生灌丛、竹丛及林缘草地常见。摄影／赵纳勋
This non-migratory bird is often seen in lower-altitude second growth shrublands, bamboo shrubs and meadows at the forest line. Photography by Zhao Naxun

7. 白鹡鸰 White Wagtail *Motacilla alba*
留鸟。体羽为黑白两色，栖息于社区村落、河流、小溪、水塘附近。摄影／赵纳勋
This non-migratory bird has a back and white plumage. They live in community grounds, villages, rivers, brooks, reservoirs and areas nearby. Photography by Zhao Naxun

8. 白颈鸦 Collared Crow *Corvus torquatus*
留鸟。栖息于低海拔河滩和村庄田野。摄影／赵纳勋
The collared crow is a non-migratory bird that inhabits lower-altitude river banks, villages and fields. Photography by Zhao Naxun

1. 白眶鸦雀 Spectacled Parrotbill *Paradoxornis conspicillatus*
留鸟。广泛分布于保护区中低海拔山地、竹林及灌丛中。摄影/张永文
This non-migratory bird is widely distributed in the nature reserve within the mid-low altitude mountainous bamboo forests and shrublands. Photography by Zhang Yongwen

2. 白领凤鹛 White-collared Yuhina *Yuhina diademata*
留鸟。成对或集小群活动于区内中高海拔灌丛,冬季活动范围下移。摄影/赵纳勋
The white-collared yuhina is a non-migratory bird species active in pairs or in small flocks. They live in the nature reserve at mid-high altitude shrublands and move to lower altitudes to winter. Photography by Zhao Naxun

3. 白眉[姬]鹟 Yellow-rumped Flycatcher *Ficedula zanthopygia*
候鸟。下体金黄与上体的主体黑色对比明显,白色的眉线及翼斑愈显其美丽。常见于中低海拔近水灌丛林地。摄影/赵纳勋
This migratory bird has golden underparts that contrast with a black upper body. Its white brow lines and wing patterns add to the beauty of this bird species, which are found in mid-low altitude shrublands and forests readily accessible to water sources. Photography by Zhao Naxun

4. 白腰文鸟 White-rumped Munia *Lonchura striata*
留鸟。常见于保护区中低海拔森林及社区林缘,为本地优势鸟种。摄影/张林
This non-migratory bird is found in the nature reserve's mid-low altitude forests, timberlines, and communities. They are a dominant bird species in the area. Photography by Zhang Lin

5、6. 白头鹎 Light-vented Bulbul *Pycnonotus sinensis*
留鸟。本地优势种,春季集群活动,多达上百只,蔚为壮观,是华阳站区一道亮丽的风景。摄影/张永文 & 赵纳勋
The light-vented bulbul is a dominant resident species in the local area. In the spring, they gather in hundreds forming an awesome spectacle, which has become a scenic appeal for Huayang ranger grounds. Photography by Zhang Yongwen & Zhao Naxun

1	2	8
3	4	9
5	6	10
7		

1. 斑背噪鹛 Barred Laughingthrush *Garrulax lunulatus*
留鸟。中国的特有种，秦岭优势种。有明显的白色眼斑，栖息于海拔较高的针叶林、针阔叶混交林及竹林、林缘地带。摄影／向定乾
This non-migratory bird is endemic to China and a dominant species in Qinling. It has prominent white marks over its eyes and inhabits in the higher altitude needleleaf forests, mixed broadleaf-needleleaf forests, bamboo forests and near timberlines. Photography by Xiang Dingqian

2. 斑鸫 Dusky Thrush *Turdus naumanni*
候鸟。体型约25厘米左右，具浅棕色翼线和棕色宽阔翼斑。常见于保护区低海拔及周边社区森林、村落和农田。摄影／张永文
The dusky thrush is a migratory bird species that measures about 25 centimeters. They have light brown rims and wide brown bands on their wings. They are found in the lower altitudes in the nature reserve, close to community-bordering forests, villages and farmlands. Photography by Zhang Yongwen

3. 白斑翅拟蜡嘴雀 White-winged Grosbeak *Mycerobas carnipes*
留鸟。见于海拔2000米以上的针叶林和混交林，以植物的种子和昆虫为食。摄影／向定乾
The resident white-winged grosbeak is found in needleleaf forests and mixed forests around 2,000 meters in altitude. Plant seeds and insects are its primary food. Photography by Xiang Dingqian

4. 宝兴歌鸫 Chinese Thrush *Turdus mupinensis*
留鸟。中国特有种。栖息于中低海拔针叶林和针阔混交林，常在林下灌丛活动觅食。摄影／赵纳勋
This non-migratory bird is endemic to China, inhabiting mid-low altitude needleleaf forests and mixed broadleaf-needleleaf forests, hunting for food in shaded shrubs. Photography by Zhao Naxun

5. 北红尾鸲（雌）Daurian Redstart (female) *Phoenicurus auroreus*
留鸟。无论雌雄翅膀上都有醒目的白色翼斑，生境类型多样，本区常见。摄影／赵纳勋
The daurian redstart is a non-migratory bird with eye-catching white patterns on its wings. They live in diverse environments and can often be seen in the nature reserve. Photography by Zhao Naxun

6. 北红尾鸲（雄）Daurian Redstart (male) *Phoenicurus auroreus*
摄影／赵纳勋 Photography by Zhao Naxun

7. 北椋鸟 Purple-backed Starling *Sturnus sturninus*
候鸟。长约16厘米，见于低海拔开阔的山地疏林中。摄影／张永文
This migratory bird is a small bird, measuring only 16 centimeters in size. They are sighted in the open areas of lower altitude mountainous spare trees. Photography by Zhang Yongwen

8. 橙翅噪鹛 Elliot's Laughingthrush *Garrulax elliotii*
留鸟。秦岭优势种，本区常见。雌雄羽色相似，外侧飞羽基部、尾羽外侧橙黄色，虹膜浅乳白色。摄影／赵纳勋
This non-migratory bird is a dominant species in Qinling and can often be seen in the nature reserve. The two sexes have similar plumage, having yellow markings on their flight feathers and tail feathers, and ivory iris. Photography by Zhao Naxun

9. 长尾山椒鸟（雌）Long-tailed Minivet (female) *Pericrocotus ethologus*
夏候鸟。常在树冠上空盘旋，羽色艳丽易于观察，秋季集大群活动于中低海拔针阔混交林。摄影／向定乾
The long-tailed minivet is a summer migrant. They are often seen hovering in the air above tree crowns. They have a rich colorful plumage readily noticeable. In the autumn large flocks become active in mid-low altitude mixed broadleaf-needleleaf forests. Photography by Xiang Dingqian

10. 长尾山椒鸟（雄）Long-tailed Minivet (male) *Pericrocotus ethologus*
摄影／向定乾 Photography by Xiang Dingqian

1. 橙头地鸫 Orange-headed Thrush *Zoothera citrina*
善于鸣唱的候鸟，鸣声甜美清晰，区内中低海拔区域偶见，常躲藏在浓密树冠覆盖下的地面。摄影／胡万新
A keen singer with a clear, pleasing voice, this migratory thrush is usually seen at mid-low altitudes, often on forest floors shaded by dense tree canopies. Photography by Hu Wanxin

2. 发冠卷尾 Spang led Drongo *Dicrurus hottentottus*
候鸟。区内广泛分布，细长的羽冠，钝而上翘的尾，具黑色金属光泽，使人过目不忘。摄影／赵纳勋
This migratory bird is widely distributed in the nature reserve. It has a non-erect, thin, long feathered crest and curvy forked tail. Its iridescent blue-black plumage makes a memorable impression. Photography by Zhao Naxun

3. 淡绿鵙鹛 Green Shrike-Babbler *Pteruthius xanthochlorus*
留鸟。区内常见，针叶林、针阔林均有分布，常与山雀、鹛和柳莺混群。摄影／张永文
This non-migratory bird is frequently sighted in the nature reserve. They are distributed in both needleleaf forests and broadleaf forests, mixing with other birds including tits, vulvettas, laughing thrushes and warblers. Photography by Zhang Yongwen

4. 方尾鹟 Grey-headed Canary-Flycat *Culicicapa ceylonensis*
本地繁殖，头灰色，上体绿下体黄。常活跃于长青中低海拔的森林。摄影／赵纳勋
Locally multiplied, grey head with green upper body and yellow lower body, commonly seen in the low altitude forest in the Changqing Nature Reserve. Photography by Zhao Naxun

5. 褐冠山雀 Grey-crested Tit *Parus dichrous*
留鸟。体小而色淡的山雀，栖息于区内海拔较高的针叶林。摄影／向定乾
This non-migratory bird is small in size and pale in color. In the nature reserve, they inhabit the needleleaf forests at higher altitudes. Photography by Xiang Dingqian

6. 橙胸[姬]鹟 Rufous-gorgeted Flycatche *Ficedula strophiata*
候鸟。体型较小的林栖型鹟，常见于不同海拔的林中灌丛。摄影／赵纳勋
A migratory bird, the rufous-gorgeted flycatcher is a small tree-dwelling species. They are found in shrublands of various altitudes. Photography by Zhao Naxun

7. 粉红胸鹨 Rosy Pipit *Anthus roseatus*
候鸟。为中等体型的偏灰色而具纵纹的鹨，繁殖期下体、眉纹粉红。活动于低海拔溪流及稻田。摄影／赵纳勋
This migrant is a medium-sized pipit with greyish patterns. During the breeding season, its underparts and brow line turn pink. They are active in lower-altitude streams and paddy fields. Photography by Zhao Naxun

8. 大山雀 Great Tit *Parus major*
留鸟。数量多，常见于多种生境。摄影／赵纳勋
This non-migratory bird is found in large numbers living in various habitats. Photography by Zhao Naxun

秦岭长青
野生生命的庇护所
CHANGQING, QINLING
A Natural Shelter to Wildlife

1. 赤颈鸫（雄）Dark-throated Thrush (male) *Turdus ruficollis*
候鸟。以昆虫和植物种子为食，开阔灌丛、疏林地常见。摄影／赵纳勋
A migratory bird, the dark-throated thrush feed on insects and plant seeds as its primary food source, and are often sighted in open shrublands and sparse trees. Photography by Zhao Naxun

2. 赤颈鸫（雌）Dark-throated Thrush (female) *Turdus ruficollis*
摄影／张永文 Photography by Zhang Yongwen

3. 大嘴乌鸦 Large-billed Crow *Corvus macrorhynchos*
留鸟。雌雄同形同色，嘴形粗大，上嘴前缘与前额几成直角。多见于村落农田。摄影／赵纳勋
The large-billed crow is a non-migratory bird with both sexes identical in physical characteristics. They have large bills that appear perpendicular to its forehead. They are notably found in village farms. Photography by Zhao Naxun

4. 高山短翅莺 Russet Bush-Warbler *Bradypterus seebohmi*
留鸟。栖息于保护区海拔2000米以上的中高海拔灌丛密林之中，体长约13厘米。摄影／向定乾
A resident bird, the russet bush-warbler inhabits the dense bushes 2,000 meters above sea level within the Changqing Nature Reserve. Its full body length is about 13 centimeters. Photography by Xiang Dingqian

5. 短趾百灵 Red-capped Lark *Calandrella brachydactyla*
候鸟。盘旋飞行，时作短促、突发性、重复性的鸣唱，见于低海拔社区田野。摄影／向定乾
This migratory lark makes short, sudden, repeated calls when hovering. They are sighted in fields and near human communities at lower-altitudes. Photography by Xiang Dingqian

6. 戈氏岩鹀 Godlewski's Bunting *Emberiza godlewskii*
候鸟。见于中低海拔干燥而多岩石的沟壑山坡，或近森林而多灌丛的沟壑深谷。摄影／赵纳勋
The migratory Godlewski's bunting can be found on dry, rocky foothills or shrub-lined gullies at mid-low altitudes. Photography by Zhao Naxun

7. 黄腹树莺 Yellowish-bellied Bush Warbler *Cettia acanthizoides*
留鸟。夏季栖息于高海拔针叶林下竹林灌丛，冬季下至海拔1000米以下活动。摄影／胡万新
A resident bird, this bush warbler lives in the bamboo shrubs at high altitude needleleaf forests during the summer. In the winter it moves downhill to regions below an altitude of 1,000 meters. Photography by Hu Wanxin

8. 冠纹柳莺 Blyth's Leaf Warbler *Phylloscopus reguloides*
候鸟。在本区繁殖，栖息于中低海拔的阔叶林及灌丛，常与其他柳莺混群活动。摄影／赵纳勋
The Blyth's leaf warbler is a migrant bird that breeds within the nature reserve, inhabiting mid-low altitude broadleaf forests and shrublands, and often mixing flocks with other warblers. Photography by Zhao Naxun

1. 黑［短脚］鹎 Madagascar Bulbul *Hypsipetes leucocephalus*
候鸟。于4、5月份飞来，本区杨家沟、吊坝河一带可见。活动于树冠，叫声喧杂。摄影／张永文
Migratory, the black bulbul visits the nature reserve in April and May. They can be seen around Yangjiagou and the Diaoba River. Usually active in tree tops, they are noisy birds. Photography by Zhang Yongwen

2. 黑尾蜡嘴雀 Yellow-billed Grosbeak *Eophona migratoria*
候鸟。嘴粗大，蜡黄色，翅和尾黑色。偶见于保护区境内。摄影／张永文
This migratory bird has a large yellow bill, black wings and tail. They are occasionally sighted within the nature reserve. Photography by Zhang Yongwen

3. 黑喉红尾鸲（雄）Hodgson's Redstart (male) *Phoenicurus hodgsoni*
留鸟。常活动于近溪流的开阔林地和灌丛。摄影／胡万新
This non-migratory bird is active near streams, open woods and shrublands. Photography by Hu Wanxin

4. 黑喉红尾鸲（雌）Hodgson's Redstart (female) *Phoenicurus hodgsoni*
摄影／胡万新 Photography by Hu Wanxin

5. 黑喉石鸲（雌）Common Stonechat (female) *Saxicola torquata*
候鸟。油菜花盛开的季节，黑喉石鸲就会光临保护区华阳、茅坪一带，活动于河谷两岸的森林和田野。摄影／赵纳勋
During the blooming season of rapeseed, the stonechats migrate to the Changqing Nature Reserve, Huayang and Maoping areas. They are active in forests and fields along the banks of river valleys. Photography by Zhao Naxun

6. 黑喉石鸲（雄）Common Stonechat (male) *Saxicola torquata*
摄影／赵纳勋 Photography by Zhao Naxun

7. 褐河乌 Brown Dipper *Cinclus pallasii*
留鸟。全身体羽深褐色，栖息于溪流河谷，以水生生物为食。摄影／赵纳勋
Dark brown throughout its body, the brown dipper inhabits streams and ravines, feeding on aquatic organisms. Photography by Zhao Naxun

8. 黑冠山雀 Rufous-vented Tit *Parus rubidiventris*
留鸟。冠羽及胸兜黑色，脸颊白，上体灰色。见于保护区中高海拔针叶林和混交林。摄影／向定乾
This resident bird has a black crest and breast, white cheeks and a grey upper body. They may be found in the nature reserve's mid-high altitude needleleaf forests and mixed leaf forests. Photography by Xiang Dingqian

9. 黑卷尾 Black Drongo *Dicrurus macrocercus*
候鸟。中等体型具辉蓝色光泽的卷尾，尾长而分叉。常见中低海拔林缘地带。摄影／赵纳勋
This medium-sized black bird is migratory and has a long, forked tail with a blue iridescent sheen. It is often seen in mid-low altitudes near timberlines. Photography by Zhao Naxun

10. 黑领噪鹛 Greater Necklaced Laughing-thrush *Garrulax pectoralis*
候鸟。体型略大的棕褐色噪鹛，胸有一黑色环带。常见集群活动于区内中低海拔针阔混交林内。摄影／胡万新
Migratory and slightly larger in size, this laughing-thrush has a black band on its breast. It is often sighted in large flocks in the nature reserve within the mid-low altitude mixed broadleaf-needleleaf forests. Photography by Hu Wanxin

1	3	5	8
2	4	6	
		7	9	10

秦岭长青 野生生命的庇护所
CHANGQING, QINLING
A Natural Shelter to Wildlife

1. 黑喉歌鸲（雄）Blackthroated (male) *Luscinia obscura*
被誉为"鸟中的大熊猫"，十分珍稀，非常罕见。长青的保护工作者于海拔2500米处发现了它，并用照片记录了它的繁殖习性。摄影／张永文
The blackthroat is regarded as the giant panda among birds for its rarity and significance. Conservation workers from Changqing discovered this species at 2,500 meters above sea level, and recorded its reproductive habits on camera. Photography by Zhang Yongwen

2. 黑喉歌鸲（雌）Blackthroated (female) *Luscinia obscura*
摄影／张永文 Photography by Zhang Yongwen

3. 红尾伯劳（雌）Brown Shrike (female) *Lanius cristatus*
夏候鸟。体长约20厘米，尾上覆羽红棕色，头顶灰色或红棕色。见于保护区及周边中低海拔地区。摄影／张永文
This summer migrant bird measures 20 centimeters in length. It has reddish brown feathers on its tail, and brown or grey on its crown. They can be found in the nature reserve at mid-low altitudes. Photography by Zhang Yongwen

4. 红尾伯劳（雄）Brown Shrike (male) *Lanius cristatus*
摄影／赵纳勋 Photography by Zhao Naxun

5. 红尾水鸲（雄）Plumbeous Water-redstart (male) *Rhyacornis fuliginosus*
留鸟。体小约14厘米，雌雄异色，河谷溪流旁常见。摄影／赵纳勋
The plumbeous water-redstart is a small non-migratory bird measuring roughly 14 centimeters. The two sexes are different in color. They are regularly seen along ravines or streams. Photography by Zhao Naxun

6. 红尾水鸲（雌）Plumbeous Water-Redstart (female) *Rhyacornis fuliginosus*
摄影／赵纳勋 Photography by Zhao Naxun

7. 红胁蓝尾鸲 Orange-flanked Bush-Robin *Tarsiger cyanurus*
候鸟。雌雄都具有红棕色的胁部和蓝色的尾巴。分布于保护区及周边中低海拔区域。摄影／赵纳勋
The organe-flanked bush-Robin is a migratory bird. Both male and female birds have reddish flanks and blue tails. They are distributed in the mid-low altitudes in and around the nature reserve. Photography by Zhao Naxun

1. 褐头鹪莺 Plain Prinia *Prinia inornata*
留鸟。活动范围广，常见集小群鸣叫飞行于树干、草茎间。摄影／向定乾
The plain prinia is a non-migratory bird active in a wide range of territories. They are often seen in small flocks calling and flying from tree to tree or in the grasses. Photography by Xiang Dingqian

2. 红头穗鹛 Rufous-capped Babbler *Stachyris ruficeps*
留鸟。顶冠棕色，上体暗灰橄榄色，喉具黑色细纹。常见栖于森林、灌丛及竹丛。摄影／胡万新
This non-migratory bird has a brown crest, dark olive upper parts and black lines on its throat. They are often sighted in forests, shrublands and bamboo forests. Photography by Hu Wanxin

1	2	7
3	4	8
5	6	9

野生生命的庇护所
CHANGQING, QINLING
A Natural Shelter to Wildlife

3. 红翅旋壁雀 Wallcreeper *Tichodroma muraria*
留鸟。飞行姿态优雅，翅膀绯红色斑纹使之色彩鲜艳，常见于保护区内沟谷悬崖峭壁处。摄影／赵纳勋
This non-migratory bird exhibits an elegant form while in flight. The pink patterns covering its wings make a colorful display. They are found in the nature reserve in the cliffs along canyons. Photography by Zhao Naxun

4. 赭红尾鸲 Black Redstart *Phoenicurus ochruros*
留鸟，常见于开阔区域的林缘、社区和农田。摄影／赵纳勋
Resident bird, often seen along forest lines around open fields, in communities and farms. Photography by Zhao Naxun

5. 红腹山雀 Rusty-breasted Tit *Parus davidi*
留鸟。中国特有种。栖息于保护区内海拔较高的阔叶林、针阔混交林及针叶林内，常活动于树冠。摄影／张永文
Endemic to China, this non-migratory bird resides in the nature reserve's higher altitude broadleaf forests, mixed broadleaf-needleleaf forests and needleleaf forests. They are usually active in the crowns of trees. Photography by Zhang Yongwen

6. 红头[长尾]山雀 Black-throated Tit *Aegithalos concinnus*
留鸟。体小约10厘米。特征明显，性情活泼，常成群活动于山地森林和灌木林间。摄影／赵纳勋
This small non-migratory tit measures 10 centimeters long. They are distinctive, lively and active in flocks. They can be found in mountainous forests and shrublands. Photography by Zhao Naxun

7. 褐头雀鹛 Streak-throated Fulvetta *Alcippe cinereiceps*
留鸟。中等体型（12厘米）的褐色雀鹛，虹膜白色，分布于区内中高海拔区。摄影／赵纳勋
This non-migratory bird is a medium-sized (12 centimeters) fulvetta with brown plumage and white irises. They live in the mid-high altitudes of the reserve. Photography by Zhao Naxun

8. 黑枕黄鹂 Black-naped Oriole *Oriolus chinensis*
候鸟。俗称黄鹂。区内外中低海拔广泛分布，于5、6月份在本地繁殖。摄影／赵纳勋
A migratory bird, the black-naped oriole is widely distributed throughout the nature reserve, in mid-low altitudes. They come to the local area in May and June for breeding. Photography by Zhao Naxun

9. 红交嘴雀 Red Crossbill *Loxia curvirostra*
留鸟。亦称"歪嘴雀"，独特交叉的嘴有利于从松果中取食种子。常见于高海拔针叶林。摄影／赵纳勋
A non-migratory bird, the red crossbill has a unique crossed bill to help the extraction of conifer nuts from their cones. The red crossbills are found in high-altitude needleleaf forests. Photography by Zhao Naxun

野生生命的庇护所
CHANGQING, QINLING
A Natural Shelter to Wildlife

1. 红嘴蓝鹊 Red-billed Blue Magpie *Urocissa erythrorhyncha*
留鸟。能发出多种叫声，为较凶悍的杂食性鸣禽，广泛分布，随处可见。摄影／赵纳勋
The red-billed blue magpie is a non-migratory bird able to give out a variety of calls. It is an aggressive omnivorous songbird frequently seen throughout the nature reserve. Photography by Zhao Naxun

2. 虎斑地鸫 Scaly Thrush *Zoothera dauma*
候鸟。为具粗大褐色鳞状斑纹的地鸫，见于保护区中低海拔地区，于森林地面觅食。摄影／张永文
The scaly thrust is a migratory bird which has thick brown scaly patterns on its body. They may be found at mid-low altitudes of the nature reserve. They are adapted to hunting for food on the forest grounds. Photography by Zhang Yongwen

3. 黄腹山雀 Yellow-bellied Tit *Parus venustulus*
留鸟。中国特有种。体型较小，保护区中低海拔森林内常见。摄影／赵纳勋
This small resident bird is endemic to China and often found in mid-low altitude forests. Photography by Zhao Naxun

4. 戴菊 Goldcrest *Regulus regulus*
留鸟，体型较小，主要特征为顶冠纹金黄色或橙红色（雄），两侧并以黑色冠纹。见于区内杨家沟、吊坝河。摄影／华英
The goldcrest is a resident bird, small in body size, with a golden or orange crest male, and black stripes. Commonly seen in Yangjiagou and Diaobahe in the reserve. Photography by Hua Ying

5. 红喉歌鸲 Siberian Rubythroat *Luscinia calliope*
候鸟。成年雄鸟的特征为喉红色，常活动于近溪流的林缘灌丛、草丛和菜地。摄影／华英
The Siberian rubythroat is a resident bird commonly seen near waters, bushes, underbrush and vegetable fields. The adult male bird has a red throat. Photography by Hua Ying

6. 红胁绣眼鸟 Chestnut-flanked White-eye *Zosterops erythropleurus*
候鸟。与暗绿绣眼鸟最大的区别在于两胁栗色，二者常混群活动，见于保护区及周边中低海拔区。摄影／赵纳勋
The distinction between this migratory white-eye and the Japanese white-eye is the chestnut color in its flanks. The flocks of these two species often mix, and they can be seen in the mid-low altitude areas in the nature reserve and its surroundings. Photography by Zhao Naxun

7. 红嘴鸦雀 Great Parrotbill *Conostoma aemodium*
留鸟。本区优势种。为体型大的褐色鸦雀，叫声清晰而富韵味，常见于区内中高海拔秦岭箭竹林缘。摄影／赵纳勋
The great parrotbill is a dominant resident species in the area. It is a large brown bird that is capable of making clear and charming calls. They are found in the nature reserve's mid-high altitudes, along the timberlines of Qinling's umbrella bamboo forests. Photography by Zhao Naxun

秦岭长青 野生生命的庇护所
CHANGQING, QINLING
A Natural Shelter to Wildlife

```
1 3 | 7 8
  4
  5
2 6
```

1. 红嘴相思鸟 Red-billed Leiothrix *Leiothrix lutea*
留鸟。又称相思鸟，本地常见种。羽色艳丽、鸣声婉转。摄影／胡万新
This non-migratory bird is also known as the Pekin nightingale. It is frequently sighted in the local area and has colorful plumage and a beautiful voice. Photography by Hu Wanxin

2. 画眉 Hwamei *Garrulax canorus*
留鸟。鸣声悦耳洪亮，婉转动听。常见于保护区中低海拔及周边林区。摄影／赵纳勋
The hwamei is a non-migratory bird with a loud and pleasing singing voice. They are found in the nature reserve's mid-low altitudes and surrounding forests. Photography by Zhao Naxun

3. 黄喉鹀 Yellow-throated Bunting *Emberiza elegans*
留鸟。雌雄体色有别，数量较多，常见于区内较开阔的山林河谷地带。摄影／赵纳勋
The yellow-throated bunting is a resident bird. The male and female birds bear different plumage. Greater in number, they are found by the ravines of more open forests. Photography by Zhao Naxun

4. 黄喉鹀（雌）Yellow-throated Bunting (female) *Emberiza elegans*
摄影／赵纳勋 Photography by Zhao Naxun

5. 虎纹伯劳 Tiger Shrike *Lanius tigrinus*
候鸟。顶冠及颈背灰色，背、两翼及尾浓栗色而多具黑色横斑，过眼线宽且黑。活动于多林地带。摄影／赵纳勋
This migratory bird has a grey crown and nape. Their wings and tail are dark chestnut brown, often with black horizontal patterns. They have a thick black mask over each eye and are active in densely wooded forests. Photography by Zhao Naxun

6. 黄臀鹎 Brown-breasted Bulbul *Pycnonotus xanthorrhous*
留鸟。种群大，数量多，分布广，极常见。摄影／赵纳勋
Brown-breasted bulbuls are non-migratory with large local populations and a wide distribution in the area. Photography by Zhao Naxun

7. 黄头鹡鸰 Citrine Wagtail *Motacilla citreola*
候鸟。其觅食的倩影常见于保护区及周边中低海拔河流、溪谷、湿地。摄影／张永文
These migrant wagtails can be seen hunting for food in the nature reserve and nearby mid-low altitude rivers, ravines and wetlands. Photography by Zhang Yongwen

8. 黄鹡鸰 Yello Wagtail *Motacilla Wagtail*
候鸟。头灰背绿，当春暖花开时就会光临保护区低海拔的溪流河谷。摄影／赵纳勋
Migratory bird, grey head with green back, commonly seen in the river areas of the nature reserve in spring. Photography by Zhao Naxun

1. 黄额鸦雀 Fuivous Parrotbill *Paradoxornis fulvifrons*
留鸟。常见于区内中高海拔针阔混交林下竹林及灌丛中，喜集群活动。摄影／张永文
The fuivous parrotbill is a resident bird found in mixed broadleaf-needleleaf forests, bamboo forests and shrublands. They are a gregarions bird species. Photography by Zhang Yongwen

2. 灰鹡鸰 Grey Wagtail *Motacilla cinerea*
候鸟。灰色的上体和黄色的下体型成鲜明对照。中低海拔的溪流河谷地常见。摄影／胡万新
This migratory bird has contrasting grey upperparts and yellow underparts. They are often sighted by the mid-low altitude streams and ravines. Photography by Hu Wanxin

3. 灰翅噪鹛 Moustached Laughing-thrush *Garrulax cineraceus*
留鸟。雌雄羽色相似，常见小群活动于海拔 1800 米以下的巴山木竹林和灌丛。摄影／赵纳勋
The male and female birds of this resident species have similar plumage. They are often seen in small flocks, active in *Bashania fargesii* bamboo forests and shrublands at or below 1,800 meters in altitude. Photography by Zhao Naxun

4. 灰背伯劳 Grey-backed Shrike *Lanius tephronotus*
候鸟。头顶至下背暗灰，区内外灌丛开阔地常见。摄影／赵纳勋
A migratory bird, the grey-backed shrike has grey plumage on its head and through its back. They are often seen outside of the nature reserve near shrublands in open spaces. Photography by Zhao Naxun

5. 灰蓝 [姬] 鹟 Slaty-blue Flycatcher *Ficedula tricolor*
候鸟。体小，约 13 厘米，为青石蓝色鹟，偶见于保护区及周边中低海拔林下灌丛。摄影／胡万新
These small migratory birds measure 13 centimeters in length and has stone blue plumage throughout their back. They are occasionally sighted in shaded shrublands at mid-low altitudes. Photography by Hu Wanxin

6. 灰林䳭（雄）Grey Bushchat (male) *Saxicola ferrea*
留鸟。常见于开阔林间的灌丛顶端。摄影／赵纳勋
This non-migratory bird is found in bush tops in open wooded areas. Photography by Zhao Naxun

7. 灰林䳭（雌）Grey Bushchat (female) *Saxicola ferrea* 摄影／胡万新
Photography by Hu Wanxin

8. 灰卷尾 Ashy Drongo *Dicrurus leucophaeus*
候鸟。中等体型的灰色卷尾，脸偏白，尾长而开叉。见于保护区及周边中低海拔林缘。摄影／赵纳勋
A migratory, medium-sized drongo, this bird has a palish face and long forked tail. They may be found in the nature reserve or its surrounding mid-low altitude tree lines. Photography by Zhao Naxun

9. 灰冠鹟莺 Grey-crowned Warbler *Seicercus tephrocephalus*
候鸟。头顶灰色纹，无翼斑。多隐匿于林下层，叫声嘹亮。摄影／向定乾
This migratory bird has grey patterns on its crown and no markings on the wing. They hide under vegetation and chirps with vigor. Photography by Xiang Dingqian

野生生命的庇护所
CHANGQING, QINLING
A Natural Shelter to Wildlife

1. 灰头灰雀（雄）Grey-headed Bullfinch (male) *Pyrrhula erythaca*
留鸟。为体型略大（17厘米）而壮实的灰雀。保护区苍耳崖、柏杨坪一带常见，喜集群活动。摄影／张永文
This non-migrant bird is a slightly larger (17 centimeters) and heavier kind of bullfinch found near the areas of Cangerya and Boyangping. They are often seen in flocks. Photography by Zhang Yongwen

2. 灰头灰雀（雌）Grey-headed Bullfinch (female) *Pyrrhula erythaca*
摄影／赵纳勋 Photography by Zhao Naxun

3. 灰眶雀鹛 Grey-cheeked Fulvetta *Alcippe morrisonia*
留鸟。鸣声甜美，栖息于中低海拔林下矮树和灌丛。摄影／赵纳勋
The grey-cheeked fulvetta is a non-migratory bird with a sweet singing voice. It inhabits short trees and shrubs located at mid-low altitudes. Photography by Zhao Naxun

4. 灰头鸫 Grey-headed Thrush *Turdus rubrocanus*
留鸟。体长约23厘米，且多栗色。常于树顶鸣叫，声音淳美。摄影／胡万新
The grey-headed thrush is a non-migratory bird which measures 23 centimeters approximately. They have chestnut brown plumage and are usually heard in tree tops singing with a pleasing voice. Photography by Hu Wanxin

5. 灰头鹀 Black-faced Bunting *Emberiza spodocephala*
候鸟。为体小的黑色及黄色鹀，常见于区内低海拔地区。摄影／赵纳勋
This migratory bird is small in size, with black and yellow plumage. They are found in lower altitudes. Photography by Zhao Naxun

6. 灰翅鸫 Grey-winged Blackbird *Turdus boulboul*
体长约28厘米，2012年5月首次发现拍摄于长青保护区吊坝河海拔1400米的阔叶林内。摄影／张永文
Approximately 28 centimeters in length, the grey-winged blackbird was first discovered and photographed in the broadleaf forests at 1,400 meters in altitude in the reserve. Photography by Zhang Yongwen

秦岭长青
野生生命的庇护所
CHANGQING, QINLING
A Natural Shelter to Wildlife

```
  2 | 5 6
1 3 |
  4 | 7
```

1. 火冠雀（雄）Fire-capped Tit (male) *Cephalopyrus flammiceps*
候鸟。每年 4 月，当区内中低海拔地区柳树、杨树开花时，火冠雀就会到来，花谢之时会迁徙而去。摄影／向定乾
The fire-capped tit is a migratory bird which arrives in the reserve annually in April, when willow trees and poplar trees at mid-low altitudes are in bloom. It departs when the blooms are over. Photography by Xiang Dingqian

2. 火冠雀（雌）Fire-capped Tit (female) *Cephalopyrus flammiceps*
摄影／向定乾 Photography by Xiang Dingqian

3. 家燕 Barn Swallow *Hirundo rustica*
候鸟。初春的保护区华阳、茅坪等地，常能看到家燕低空盘旋于水面追逐捕食的倩影。摄影／赵纳勋
The migratory bird comes to Huayang and Maoping areas in early spring. They are often seen hovering over water surfaces hunting for food. Photography by Zhao Naxun

4. 鹪鹩 Wren *Troglodytes troglodytes*
留鸟。体小巧短胖，尾短而上翘，仅作短距离飞行。近水的灌丛、林下常见。摄影／赵纳勋
The wren is a small resident bird with a rounded body and uplifted tail. They only fly short distances and can often be seen in shrubs and in tree shades near the water. Photography by Zhao Naxun

5. 灰椋鸟 White-cheeked Starling *Sturnus cineraceus*
留鸟。本地常见种，体型与丝光椋鸟接近，但体色偏褐色，头部色深，脸侧白色。摄影／赵纳勋
The white-cheeked starling is a resident bird often found in the nature reserve grounds. It resembles the silky stanling in shape and has brown plumage darker towards its head and white cheeks. Photography by Zhao Naxun

6，7. 金翅 [雀] Grey-capped Greenfinch *Carduelis sinica*
留鸟。本区优势种，区内林地边缘、灌丛和荒野田地常见。摄影／赵纳勋 & 张永文
This non-migratory bird is a dominant species that can be seen in the fringes of forests, shrublands and open fields. Photography by Zhao Naxun & Zhang Yongwen

秦岭长青 野生生命的庇护所
CHANGQING, QINLING
A Natural Shelter to Wildlife

1. 极北柳莺 Arctic Warbler *Phylloscopus borealis*
候鸟。具黄白色长眉纹，上体深橄榄色及甚浅的白色翼斑。区内中低海拔原始林及次生林常见。摄影／胡万新
A migrant bird with yellow and white feathers, and a dark olive green upper body, the arctic warbler is commonly seen in the primary and secondary forests at middle and lower altitudes. Photography by Hu Wanxin

2. 金眶鹟莺 Golden-spectacled Warbler *Seicercus burkii*
候鸟。眼圈金黄色为其主要特征，具宽阔的绿灰色顶纹，侧缘接黑色眉纹。多活动于中高海拔林木下层。摄影／张永文
The golden-spectacled warbler is a migrant with a signature golden ring around its eye, greyish green bands on its head connected on the sides to its black brows. They are active in tree shades at mid-high altitudes. Photography by Zhang Yongwen

3. 蓝鹀（雄）Slaty Bunting (male) *Latoucheornis siemsseni*
候鸟。冬春季节常见，活动于区内海拔约1000米的近水灌丛林地。摄影／赵纳勋
A migratory bird, the slaty bunting is often sighted in spring and winter. They are active in riverside shrublands or forests around 1,000 meters in elevation. Photography by Zhao Naxun

4. 蓝鹀（雌）Slaty Bunting (female) *Latoucheornis siemsseni*
摄影／向定乾 Photography by Xiang Dingqian

5. 金色林鸲 Golden Bush-Robin *Tarsiger chrysaeus*
候鸟。夏季繁殖于保护区中高海拔松花竹林和针叶林，数量较多。摄影／赵纳勋
This migratory bird reproduces in the nature reserve's mid-high altitude bamboo forests and needleleaf forests. Large numbers of them can be seen in the area. Photography by Zhao Naxun

6. 金胸雀鹛 Golden-breasted Fulvetta *Alcippe chrysotis*
留鸟。色彩鲜艳的雀鹛，下体黄色，头偏黑，耳羽灰白。活动于区内中高海拔竹林、灌丛。摄影／胡万新
Non-migratory, the golden-breasted fulvetta is a colorful bird with yellow underparts, dark grey head and paler feathers around the ears. They are active in the area's mid-high altitude bamboo forests and shrublands. Photography by Hu Wanxin

7. 金腰燕 Red-rumped Swallow *Hirundo daurica*
夏候鸟。本地常见，浅栗色的腰与深蓝色上体成鲜明对比。摄影／赵纳勋
The red-rumped swallow is a summer migrant often seen in the area. They have contrasting light chestnut rump plumage and dark blue upper parts. Photography by Zhao Naxun

8. 酒红朱雀（雄）Vinaceous Rosefinch (male) *Carpodacus vinaceus*
留鸟。雄鸟全身深绯红色，腰色较淡，野外醒目易辨。常见于区内中高海拔竹林和灌丛。摄影／赵纳勋
A non-migratory bird, the male vinaceous rosefinch has crimson plumage throughout its body with a slightly paler rump, easily noticeable in the wild. They are mostly found in the nature reserve's mid-high altitude bamboo forests and shrublands. Photography by Zhao Naxun

9. 酒红朱雀（雌）Vinaceous Rosefinch (female) *Carpodacus vinaceus*
摄影／赵纳勋 Photography by Zhao Naxun

1. 蓝额红尾鸲（雌）Blue-fronted Redstart (female) *Phoenicurus frontalis*
留鸟。色彩艳丽的红尾鸲，雌雄尾部均有特殊的"T"型黑色图纹。区内数量多、分布广。摄影／向定乾
The blue-fronted redstart is a non-migrant bird with bright colors. Both sexes of this bird species have the unique, black T-shaped marks on their tails. They are densely distributed in the nature reserve. Photography by Xiang Dingqian

2. 蓝额红尾鸲（雄）Blue-fronted Redstart (male) *Phoenicurus frontalis*
摄影／赵纳勋 Photography by Zhao Naxun

3. 领岩鹨 Cllared Accentor *Prunella collaris*
留鸟。比较安静，不引人注意，夏季活动于保护区内中高海拔地，冬季则下移。摄影／向定乾
The alpine accentor is a quiet resident bird not often noticed. They become active in the summer in the nature reserve's mid-high altitude areas, but migrate downhill in the winter. Photography by Xiang Dingqian

4. 树鹨 Orienfnl Pipit *Anthus hodgsoni*
候鸟。活动于多种生境，从低海拔丘陵沟壑区到区内中高海拔林地都能见到其栖息的踪迹。摄影／赵纳勋
This migratory bird is active in various habitats. They can be seen both in lower-altitude foothill gullies and in the mid-high altitude forests of the nature reserve. Photography by Zhao Naxun

5. 蓝矶鸫 Blue Rock-Thrush *Monticola solitarius*
留鸟，常见于低海拔河谷近水潭突兀的岩石、房顶。鸣声悦耳，羽色独特。摄影／赵纳勋
The blue rock-thrush is a non-migratory bird found near lower-altitude ravines, on rugged rocks or rooftops. They can make pleasant vocal sounds and have a distinctive plumage. Photography by Zhao Naxun

6. 领雀嘴鹎 Collared Finchbill *Spizixos semitorques*
留鸟。区内外中低海拔广泛分布，常聚群活动。摄影／赵纳勋
The collared finchbill is a resident bird living in the mid-low altitudes outside of the nature reserve. They are widely distributed and active in flocks. Photography by Zhao Naxun

7. 栗背岩鹨 Maroon-backed Accentor *Prunella immaculata*
留鸟。为体型较小（14厘米）岩鹨。冬季活动于低海拔开阔林地，夏季栖息于高山。摄影／赵纳勋
This non-migratory bird is small in size (14 centimeter). In the winter, they are active in the lower-altitude open forests. In the summer they move to higher elevations. Photography by Zhao Naxun

8. 蓝喉太阳鸟（雄）Mrs Gould's Sunbird (male) *Aethopyga gouldiae*
秦岭山区颜色最为艳丽的一种小鸟，俊俏迷人，春季常活动于花丛之中，尤其喜食桐花。摄影／赵纳勋
The most vibrantly colored bird in the Qinling Mountains, this small bird is handsome and captivating. In the spring, it is often seen active in flower bushes. Its favorite is river mangrove flowers. Photography by Zhao Naxun

9. 蓝喉太阳鸟（雌）Mrs Gould's Sunbird (female) *Aethopyga gouldiae*
摄影／赵纳勋 Photography by Zhao Naxun

1. 普通䴓 Eurasion Nuthatch *Sitta europaea*
留鸟。体型短圆、蓝灰，腹部淡黄，两胁浓栗。攀行于树干上下啄食树皮下的昆虫。摄影／赵纳勋
The Eurasian nuthatch is a resident bird short and petite in form, with a slate back, yellow underparts, and chestnut colored flanks. It is adapted to climbing up and down on tree trunks pecking tree barks for insects underneath. Photography by Zhao Naxun

2. 栗头鹟莺 Chestnut-crowned Warbler *Seicercus castaniceps*
候鸟。体长约9厘米。春夏季常见于保护区中低海拔森林，于小树的树冠顶层觅食。摄影／张永文
Chestnut-crowned warblers are migratory birds measuring about 9 centimeters in size. They can be seen in spring and summer time hunting for food on tree tops, usually in mid-low altitude forests within the reserve. Photography by Zhang Yongwen

3. 煤山雀 Coal Tit *Parus ater*
留鸟。体小，约11厘米，具冠羽，常见于保护区海拔2400米左右的针叶林。摄影／赵纳勋
A resident bird, the full body length of the coal tit is about 11 centimeters, with crown feathers. They inhabit the coniferous forests at 2,400 meters above sea level in the Changqing Nature Reserve. Photography by Zhao Naxun

4. 冕柳莺 Eastern Crowned Warbler *Phylloscopus coronatus*
留鸟。为中等体型的柳莺，区内广泛分布，常见于林缘较大树木的树冠层。摄影／赵纳勋
Eastern crowned warblers are average in size and widely distributed in large trees along forest lines. Photography by Zhao Naxun

5. 牛头伯劳 Bull-headed Shrike *Lanius bucephalus*
留鸟。头顶褐色，背灰褐。区内低海拔疏林地常见。摄影／张永文
This brown-headed, grey brown-backed resident bird can often be seen in the nature reserve's lower altitude sparse trees. Photography by Zhang Yongwen

6. 绿背山雀 Green-backed Tit *Parus monticolus*
留鸟。很普通、也很常见，作为森林益鸟，它与自然保护工作者一起守护着长青的山林。摄影／赵纳勋
The green-backed tit is a resident bird commonly seen in the area. As a beneficial bird, it makes its own contribution to the conservation of the local ecosystem. Photography by Zhao Naxun

7. 矛纹草鹛 Chinese Babax *Babax lanceolatus*
留鸟。本地为矛纹草鹛指名亚种，栖于开阔的灌丛、棘丛及林下。摄影／向定乾
This local resident bird is a nominotypical subspecies of *Babax lanceolatus*. They live in open shrubs, thorn bushes and tree shades. Photography by Xiang Dingqian

8. 绿翅短脚鹎 Mountain Bulbul *Hypsipetes mcclellandii*
留鸟。数量较多，常见集群活动于区内中低海拔溪流、河畔、树林中层。摄影／张永文
The mountain bulbul is a non-migrant bird large in number in the local area. They are often seen in flocks and active in the middle of the woods along mid-low altitude streams. Photography by Zhang Yongwen

野生生命的庇护所
CHANGQING, QINLING
A Natural Shelter to Wildlife

```
      | 3 4
1  2  | 5 6
      | 7 8
```

145

	1	2	3	
	4	5	6	10
	7	8	9	

野生生命的庇护所
CHANGQING, QINLING
A Natural Shelter to Wildlife

1. 丝光椋鸟 Silky tanling *Sturnus sericeus*
留鸟。本地优势种，常结群活动于区内林缘地带。鲜红色而尖端黑色的嘴和头部丝状银白色为其主要特征。摄影／赵纳勋
Silky tanlings are a local dominant species often active in flocks near the area's forest lines. They have bright red bills with black tips and silvery white feathers lining their heads. Photography by Zhao Naxun

2. 普通朱雀（雄） Common Rosefinch (male) *Carpodacus erythrinus*
候鸟。形似麻雀，雄鸟头部至后颈呈鲜红色；雌鸟上体灰褐或橄榄褐色。季节性取食植物叶芽、嫩叶、花序、浆果和昆虫。摄影／赵纳勋
The common rosefinch is a migratory bird which resembles sparrows. The male has a bright red color from the crown to the nape and the female has grey brown or olive upper body. They feed on plant shoots, young leaves, inflorescent flowers, berries and insects according to the changing seasons. Photography by Zhao Naxun

3. 普通朱雀（雌） Common Rosefinch (female) *Carpodacus erythrinus*
摄影／赵纳勋 Photography by Zhao Naxun

4. 鹊鸲 Oriental Magpie Robin *Copsychus saularis*
留鸟。活动时尾巴时常翘起。喜开阔地觅食，林缘及社区常见。摄影／赵纳勋
This resident robin raises its tail when active. They are often sighted in open areas hunting for food, especially along forest lines and near communities. Photography by Zhao Naxun

5. 强脚树莺 Brownish-flanked Bush-Warbler *Cettia fortipes*
留鸟。体型略小的暗褐色树莺，常见鸣叫活动于浓密灌丛之中。摄影／赵纳勋
Smaller in size, these are dark brown resident birds often heard calling from dense shrubs. Photography by Zhao Naxun

6. 三道眉草鹀 Meadow Bunting *Emberiza cioides*
留鸟。体型略大的棕色鹀，栖息于区内低海拔林缘和灌丛。摄影／赵纳勋
Non-migratory and a slightly larger brown bunting, this bird resides in the nature reserve's lower-altitude forest lines and shrublands. Photography by Zhao Naxun

7. 山鹪莺 Striated Prinia *Striated Prinia*
留鸟。为体型略大且具上体深褐色纵纹的鹪莺，多栖于灌丛林地。摄影／张永文
These birds, non-migratory, have larger upper bodies covered in brown horizontal patterns and dwell in shrublands. Photography by Zhang Yongwen

8. 山麻雀（雄） Russet Sparrow (male) *Rasser rutilans*
留鸟。杂食性鸟类，广泛分布，喜集群活动。摄影／赵纳勋
This omnivorous, non-migratory, speciees is widely distributed and prefer group life. Photography by Zhao Naxun

9. 山麻雀（雌） Russet Sparrow (female) *Rasser rutilans*
摄影／赵纳勋 Photography by Zhao Naxun

10. 三趾鸦雀 Three-toed Parrotbill *Paradoxornis paradoxus*
留鸟。中国特有种，秦岭是其主要分布区。羽色似白眶鸦雀，但体型较大、三趾为主要特征。栖息于区内中高海拔针叶林、竹林和灌丛中。摄影／向定乾
The three-toed parrotbill is a resident bird endemic to China and resides primarily in Qinling. It resembles the spectacled parrotbill in color but is larger in size. Unlike other song birds, it has only three toes. They inhabit mid-high altitude needleleaf forests, bamboo forests and shrublands. Photography by XiangDingqian

1. 寿带鸟（雌）Asian Paradise-flycatcher (female) *Terpsiphone paradise*
长青保护区所在地洋县堪称寿带之乡。繁殖期，自保护区海拔1200米以下至汉江河谷，寿带鸟数量依次增多，特别是汉江河谷地带丘陵区随处可见。摄影／赵纳勋
Yangxian, where the Changqing Nature Reserve is located, is regarded as the hometown to the Asian paradise-flycatchers. During the breeding season, their numbers increase progressively with decreasing altitudes, starting from 1,200 meters above sea level within the reserve towards the Han River valley below, concentrated especially around the valley's foothills. Photography by Zhao Naxun

2. 寿带鸟（雄，栗色型）Asian Paradise-flycatcher (male, rufous morph) *Terpsiphone paradise*
摄影／赵纳勋 Photography by Zhao Naxun

3. 寿带鸟（白色型）Asian Paradise-flycatcher (white morph) *Terpsiphone paradise*
摄影／赵纳勋 Photography by Zhao Naxun

4 乌鸫 Eurasian Blackbird *Turdus merula*
留鸟。雌鸟较雄鸟色淡。稀疏林地及林缘常见，食性较杂。摄影／赵纳勋
The Eurasian blackbird is non-migratory and omnivorous. The female birds are paler in color. They are often sighted in sparse trees and along forest lines. Photography by Zhao Naxun

5. 树麻雀 Eurasian Tree Sparrow *Passer montanus*
留鸟。脸颊具明显黑色点斑。低海拔林缘和社区常见，多集群活动。摄影 / 赵纳勋
A resident bird, the Eurasian tree sparrow has obvious black spots on its cheeks. They can be seen in flocks at lower-altitude forest lines and nearby communities. Photography by Zhao Naxun

6. 水鹨 Water Pipit *Anthus spinoletta*
候鸟。为中等体型灰褐色有纵纹的鹨，繁殖期下体橙黄色。见于溪流湿地和周边稻田。摄影 / 赵纳勋
A medium-sized migratory bird, the water pipit has grey brown plumage with horizontal patterns. Its underpart turns yellow during the breeding season. They can be found in streams, wetlands and surrounding paddy fields. Photography by Zhao Naxun

7. 山鹡鸰 Forest Wagtail *Dendronanthus indicus*
夏候鸟。停栖时，尾轻轻往两侧摆动，常在林间捕食昆虫，见于区内中低海拔林缘空地和社区村落。摄影 / 赵纳勋
This summer migrant wiggles its tail when perching. They prey on insects and are found in the nature reserve's mid-low altitude forest lines and around villages. Photography by Zhao Naxun

8. 松鸦 Eurasian Jay *Garrulus glandarius*
留鸟。叫声沙哑，常在林间急促飞行，于地面和树上取食，食性较杂。区内外广泛分布。摄影 / 赵纳勋
The Eurasian jay is a non-migratory bird capable of making hoarse calls. They can be seen flying in full speed in the woods and hunting on forest floors. They are omnivorous and widely distributed in the reserve and beyond. Photography by Zhao Naxun

1. 小燕尾 Little Forktail *Enicurus scouleri*
留鸟。离巢的小燕尾缺乏觅食技巧，还需要母亲的悉心照顾。摄影／胡万新
The non-migratory little forktail lacks hunting skills upon leaving the nest, still needing its mother's care. Photography by Hu Wanxin

2. 小鳞胸鹪鹛 Pygmy Wren Babbler *Pnoepyga pusilla*
留鸟。一般生活于区内中高海拔山林以及稠密灌木丛或竹林下。摄影／张永文
The pygmy wren babbler is a resident bird living in mid-high altitude montane forests and densely wooded shrubs or bamboo forests. Photography by Zhang Yongwen

3. 小云雀 Oriental Skylark *Alauda gulgula*
留鸟。常见于区内林间的开阔草地。摄影／张永文
These birds, non-migratory, are found in forested open meadows within the nature reserve. Photography by Zhang Yongwen

4. 小鹀 Little Bunting *Emberiza pusilla*
候鸟。体小约13厘米，雌雄同色，栖息于多样生境，低海拔区常见。摄影／向定乾
This small migrant bird measures 13 centimeters in length. The two sexes bear the same color. They inhabit a variety of habitats and are usually seen at lower altitudes. Photography by Xiang Dingqian

5. 喜鹊 Black-billed Magpie *Pica pica*
留鸟。区内外中低海拔广泛分布，为本地常见种。摄影／赵纳勋
This magpie is a resident bird seen commonly throughout the nature reserve. Photography by Zhao Naxun

6. 星鸦 Spotted Nutcracker *Nucifraga caryocatactes*
留鸟。本地优势种，区内中高海拔针叶林带常见。摄影／赵纳勋
The spotted nutcracker is a resident dominant species often sighted in the mid-high altitude needleleaf forests within the reserve. Photography by Zhao Naxun

7. 烟腹毛脚燕 Asian House Martin *Delichon dasypus*
夏候鸟。栖息于保护区中低海拔区域，常成群活动于地形较开阔地带，以昆虫为食。摄影／胡万新
This summer migrant bird inhabits the mid-low altitudes in the nature reserve. They are gregarious and prefer relatively open areas and landscapes. Insects are their primary diet. Photography by Hu Wanxin

8. 小灰山椒鸟 Swinhoe's Minivet *Pericrocotus cantonensis*
候鸟。为体小的黑、灰及白色山椒鸟。区内低海拔林缘地带常见。摄影／赵纳勋
The Swinhoe's minivet is a small migrant with black, grey and white colors in its plumage. They are often sighted at the forest lines in the nature reserve's lower altitudes. Photography by Zhao Naxun

9. 小太平鸟 Japanese Waxwing *Bombycilla japonica*
候鸟，冬春季见于区内低海拔及周边林区，以植物的果实和种子为食。摄影／赵纳勋
Migrant bird, commonly seen at low altitude forest area, mainly fed on fruit and seed of local plant. Photography by Zhao Naxun

野生生命的庇护所
CHANGQING, QINLING
A Natural Shelter to Wildlife

1 2 | 3 4
    | 5 6
    | 7 8
    | 9

1. 崖沙燕 Sand Martin *Riparia riparia*
留鸟。燕子中最小的一种，栖息于河流附近沟壑陡壁、山地岩石及土崖上。摄影／胡万新
Sand martins are the smallest swallows. They dwell in gully walls, montane rocks and dirt cliffs near rivers. Photography by Hu Wanxin

2. 燕雀 Brambling *Fringilla montifringilla*
冬候鸟。胸棕腰白。冬季的华阳站区常见集大群翱翔于空中、栖息于枝头或觅食于田野。摄影／赵纳勋
A winter migrant, bramblings have brown breasts with white rumps. In winter, they can be seen flying in flocks above the Huayang ranger station, perching on tree branches or hunting for food in the fields. Photography by Zhao Naxun

3. 棕颈钩嘴鹛 Streak-breasted Scimitar Babbler *Pomatorhinus ruficollis*
留鸟。体型较小的褐色钩嘴鹛。活泼好动，区内混交林、常绿林或有竹林的矮小次生林内常见。摄影／赵纳勋
Relatively smaller for the babbler species, this non-migrant bird is brown and very lively. They are usually seen in evergreen forests, mixed forests and second growth bamboo forests. Photography by Zhao Naxun

4. 沼泽山雀 Marsh Tit *Parus palustris*
留鸟。常见种。体小，雄雌同形同色，栖息环境多样。摄影／赵纳勋
This is a small resident bird with both sexes having the same color and form. They inhabit various habitats and are often seen in the area. Photography by Zhao Naxun

5. 银喉[长尾]山雀 Long-tailed Tit *Aegithalos caudatus*
留鸟。头顶黑色，喉部中央具银灰色斑块。活泼好动，集小群活动于树冠或灌丛顶部捕食昆虫。摄影／赵纳勋
The long-tailed tit is a resident bird with a black head and silvery marks on its throat. They are lively and energetic. They hunt for insects in small flocks, in tree crowns or bushes. Photography by Zhao Naxun

6. 银脸[长尾]山雀 Sooty Tit *Aegithalos fuliginosus*
留鸟。中国特有物种。灰色的喉与白色的上胸对比而成项纹，顶冠两侧及脸银灰。常集群活动于中低海拔林区。摄影／赵纳勋
The sooty tit is a resident endemic to China. They have white breasts with gray throats forming a collar. They live in flocks in mid-low altitude forests. Photography by Zhao Naxun

7. 紫啸鸫 Blue Whistling-Thrush *Myophonus caeruleus*
夏候鸟。于本地繁殖。栖息于多石的山间溪流沟谷，成对活动，相互追逐，边飞边鸣，声音洪亮短促。摄影／赵纳勋
A small migrant, the blue whistling-thrush breeds in the local area and lives in rocky ravines or gullies. They travel in pairs, chasing and making short, powerful calls. Photography by Zhao Naxun

8. 棕背伯劳 Long-tailed Shrike *Lanius schach*
留鸟。头顶颈背灰色或灰黑色，背腰及体侧红褐，额、眼纹、两翼及尾黑色。中低海拔开阔地常见。摄影／张希明
This resident bird has grey to dark grey plumage on its head, nape and back, reddish brown on its rump and sides, and black on its wings and tail. They are seen in mid-low altitude open spaces. Photography by Zhang Ximing

1	5
2	7
3	8
4	6

1. 棕腹柳莺 Buff-throated Warbler *Phylloscopus subaffinis*
候鸟。为中等体型（10厘米）的橄榄绿色柳莺，眉纹暗黄。自低海拔向高海拔迁移活动。摄影／赵纳勋
This migrant bird is a medium-sized (10 centimeters) olive-colored warbler with tan brows. It migrates uphill from lower altitudes. Photography by Zhao Naxun

2. 棕脸鹟莺 Rufous-faced Warbler *Abroscopus albogularis*
留鸟。体小，约10厘米。上胸沾黄，色彩亮丽。常见于常绿林及竹林密丛。摄影／赵纳勋
The rufours-faced warbler is small, measuring 10 centimeters in length. They have colorful plumage with yellow tints on the upper chest, and can be found in evergreen woods or in dense bamboo bushes. Photography by Zhao Naxun

3. 棕头鸦雀 Vinous-throated Parrotbill *Paradoxornis webbianus*
留鸟。为体型纤小的粉褐色鸦雀。常成群结队栖息于中低海拔的林缘及灌丛。摄影／赵纳勋
The vinous-throated parrotbill is a non-migratory bird with a pinkish brown plumage. They live in flocks near mid-low altitude forest lines and in shrublands. Photography by Zhao Naxun

4. 棕胸岩鹨 Rufous-breasted Accentor *Prunella strophiata*
留鸟。中等体型（16厘米）的褐色具纵纹的岩鹨。分布于区内中高海拔地区灌丛及林地。摄影／向定乾
A resident bird, this medium-sized (16 centimeters) accentor is brown with patterns. They are distributed in the nature reserve's mid-high altitude shrublands. Photography by Xiang Dingqian

5. 棕尾褐鹟 Ferruginous Flycatcher *Muscicapa ferruginea*
候鸟。为体型略小的红褐色鹟，常见于林间空地和溪流两侧。摄影／胡万新
This resident bird is a smaller flycatcher with a warm brown plumage. They are often seen in open forest grounds and along streams. Photography by Hu Wanxin

6. 棕褐短翅莺 Brown Bush-Warbler *Bradypterus luteoventris*
留鸟。栖息于区内中高海拔的灌草丛内，不易发现。摄影／张永文
Brown bush-warblers dwell in higher-altitude shrublands and are not easily sighted. Photography by Zhang Yongwen

7. 斑胸钩嘴鹛 Spot-breasted Scimitar Babbler *Pomatorhinus erythrocnemis*
留鸟。体型较大的钩嘴鹛，栖息于林下灌丛，叫声响亮而独特。摄影／赵纳勋
This resident bird is a slightly larger babbler which inhabits shrub or bushes under the forest canopy. They have a bright and unique call. Photography by Zhao Naxun

8. 棕头雀鹛 Spectacled Fulvetta *Alcippe ruficapilla*
留鸟。区内常见，4月中旬开始营巢繁殖，5月中旬小鸟离巢。摄影／赵纳勋
Spectacled fulvettas are often-sighted residents. They begin building nests for breeding in April and in May juvenile birds will be ready to leave the nests. Photography by Zhao Naxun

大杜鹃寄生 Parasitic behavior of Eurasian Cuckoo
摄影 / 赵纳勋 Photography by Zhao Naxun

攀禽

Tree-climbing Birds

攀禽活动于林中或林缘，擅长于树栖攀缘，腿脚短而弱。包括鹦形目、鹃形目、鴷形目、夜鹰目和佛法僧目鸟类。攀禽大多在洞穴中筑巢，鹃形目的杜鹃科中很多种类以寄生繁殖习性著称于世。长青保护区已发现攀禽23种，从低海拔溪流河谷到高山草甸针叶林均有分布。

Tree-climbing birds are active in forests or at timberlines. Adapted to nesting in trees and climbing, they have short, delicate legs. Tree-climbing birds include the bird orders Psittaciformes, Cuculiformes, Piciformes, Caprimulgiformes and Coraciiformes. Most species of tree-climbing birds build nests in caves. Under the order Cuculiformes, many species belonging to the cuckoos are known for reproducing through brood parasitism. By 2012, 23 species of tree-climbing bird have been found to dwell in the Changqing Nature Reserve, from habitats in lower-altitude streams, canyons, to meadows and needleleaf forests.

1

2　3　4

1. 赤胸啄木鸟 Crimson-breasted Woodpecker *Dendrocopos cathpharius*
留鸟。保护区内较为少见，常栖居于阔叶林及针阔混交林内，以昆虫及花蜜为食。摄影／张永文
The crimson-breasted woodpecker is a non-migratory bird less frequently sighted in the nature reserve. They dwell in broadleaf forests and mixed broadleaf-needleleaf forests feeding on insects and honey as their primary diet. Photography by Zhang Yongwen

2. 大斑啄木鸟 Great Spotted Woodpecker *Dendrocopos major*
留鸟。保护区常见种，广泛分布于区内外。摄影／向定乾
This non-migratory bird is a commonly sighted species in the area because they are widely distributed in the nature reserve and beyond. Photography by Xiang Dingqian

3. 白背啄木鸟 White-backed Woodpecker *Dendrocopos leucotos*
留鸟，特征为下背白色、雄鸟顶冠全绯红（雌鸟顶冠黑）、额白。区内中低海拔常见。摄影／赵纳勋
Resident bird, featured by white lower back, males are with red crown (females are with black crown), white forehead, commonly seen at middle and lower altitude area of the nature reserve. Photography by Zhao Naxun

4. 斑姬啄木鸟 Speckled Piculet *Picumnus innominatus*
留鸟。体小约10厘米，橄榄色背似山雀，下体多具黑点。常见于保护区及周边中低山林。摄影／张永文
The speckled piculet is a non-migratory, small bird measuring 10 centimeters in size. It has an olive green back like that of a tit, and a black-spotted belly. They are found inside the nature reserve and the surrounding montane forests. Photography by Zhang Yongwen

1. 大杜鹃 Eurasian Cuckoo *Cuculus canorus*
候鸟。又叫布谷鸟，夏季常见于保护区及周边中低海拔林区。摄影／赵纳勋
The Eurasian cuckoo is a migratory bird also known simply as cuckoo. It is found in the summer within the nature reserve and its surrounding mid-low altitude forests. Photography by Zhao Naxun

2. 蓝翡翠 Black-capped Kingfisher *Halcyon pileata*
夏候鸟。体型较大的翠鸟，蓝色的上体和红色的嘴、脚醒目。常见于区内和周边河流、塘库沿岸。摄影／赵纳勋
The black-capped kingfisher is a larger summer migrant with a blue upper body, red bill, and bright colored feet. They can generally be found in the nature reserve or along the shores of rivers and reservoirs. Photography by Zhao Naxun

3. 冠鱼狗 Crested Kingfisher *Megaceryle lugubris*
留鸟，本地优势种。栖息于保护区低海拔河谷地带，捕鱼为食，也取食河虾、河蟹。摄影／赵纳勋
The crested kingfisher is a non-migratory, dominant species of the local area. It dwells in ravines in lower-altitude areas inside the nature reserve. Its primary diet includes fish, shrimps and crabs. Photography by Zhao Naxun

4. 灰头绿啄木鸟（雄）Grey-headed Woodpecker (male) *Picus canus*
留鸟。中等体型的绿色啄木鸟，雄鸟顶冠红色。常见于保护区及周边中低海拔林区。摄影／赵纳勋
This non-migratory bird is a medium-sized, green woodpecker. The male birds have red crests. They are found within the nature reserve and in surrounding mid-low altitude forests. Photography by Zhao Naxun

5. 灰头绿啄木鸟（雌）Grey-headed Woodpecker *Picus canus*
摄影／赵纳勋 Photography by Zhao Naxun

6. 戴胜 Eurasian Hoopoe *Upupa epops*
繁殖鸟。色彩鲜明，长而尖黑的耸立型粉棕色丝状冠羽引人瞩目。常见于保护区中低海拔疏林地及社区。摄影／赵纳勋
A breeder bird, the Eurasian hoopoe is bright in color with an eye-catching, erect, pinkish brown crown of feathers on its head. They are found in the nature reserve's mid-low altitude sparse trees and community grounds. Photography by Zhao Naxun

7. 普通翠鸟 Common Kingfisher *Alcedo atthis*
留鸟。体小，约 15 厘米，上体金属般浅蓝绿色，颈侧有白色斑点，下体呈棕色。区内外河流溪谷常见。摄影／赵纳勋
This non-migratory bird is small, measuring approximately 15 centimeters in size. It has a light metallic bluish green upper body and a brown stomach, and its neck is marked with white spots on each side. They are often sighted in the rivers and ravines in and out of the nature reserve. Photography by Zhao Naxun

秦岭长青 野生生命的庇护所
CHANGQING, QINLING
A Natural Shelter to Wildlife

1. 三宝鸟 Dollarbird *Eurystomus orientalis*
候鸟。夏季于本地繁殖，常见于保护区低海拔地区及周边社区，善于飞翔中捕食昆虫。摄影／赵纳勋
The dollarbird is a migratory bird that summer in the local area for mating and reproduction. They can be seen in the nature reserve at lower altitudes and surrounding community grounds. They are skilled at catching insects for prey while in flight. Photography by Zhao Naxun

2. 旋木雀 Eurasian Tree-Creeper *Certhia familiaris*
留鸟。常见于保护区中高海拔针叶林内。伪装性的体色与树干相近，尖硬的尾羽顶住树皮支撑身体，可长时间悬伏于树干觅食。摄影／赵纳勋
The Eurasian tree-creeper is a resident bird found in the needleleaf forests at mid-high altitudes. It is camouflaged with brown patterns similar to tree trunks and has stiff tail feathers that prop against tree barks to suspend its weight while feeding. Photography by Zhao Naxun

3. 星头啄木鸟 Grey-capped Woodpecker *Dendrocopos canicapillus*
留鸟。体小具黑白色条纹的啄木鸟，保护区内分布较广。摄影／胡万新
This non-migratory bird is small in size and marked with black and white stripes. They are distributed throughout the nature reserve. Photography by Hu Wanxin

4. 四声杜鹃 Indian Cuckoo *Cuculus micropterus*
候鸟。中等体型（30厘米）的偏灰色杜鹃，见于保护区及周边中低海拔林区。摄影／向定乾
This migratory bird is medium-built (30 centimeters), grey in color, and may be found in the mid-low altitude forests surrounding the nature reserve. Photography by Xiang Dingqian

5. 普通夜鹰 Grey Nightjar *Caprimulgus jotaka*
候鸟。中等体型（28厘米）的偏灰色夜鹰，雄鸟呈枯叶色。偶见于保护区中低海拔林区。摄影／赵纳勋
The grey nightjar is a medium-sized (28 centimeters) migratory bird, and the male nightjar displays colors resembling dead leaves. Grey nightjars are occasionally sighted in the nature reserve in the mid-low altitude forests. Photography by Zhao Naxun

6. 蚁䴕 Eurasian Wryneck *Jynx torquilla*
夏候鸟。属啄木鸟科，又称"地啄木"，体小的灰褐色啄木鸟。栖于树枝而又不攀爬，通常在地面觅食。见于保护区低海拔区域。摄影／孙承骞
Also known locally as ground peckers, the Eurasian wryneck is a small grey brown bird living on tree branches but does not climb. It is adapted to hunting for food on the forest ground and migrates in the summer. It can be found in the lower altitude areas in the nature reserve. Photography by Sun Chengqian

7. 噪鹃 Asian Koel *Eudynamys scolopacea*
候鸟。体型较大的黑色（雄鸟）杜鹃，虹膜红色。夏初常见于保护区中低海拔林区，能听到其响亮的叫声。摄影／赵纳勋
Migratory, the Asian koel is a larger (male) cuckoo with black plumage and crimson iris. During early summer, they can be found in the mid-low altitude forests where their resounding calls can be heard. Photography by Zhao Naxun

8. 高山旋木雀 Bar-tailed Tree Creeper *Bar-tailed Tree-Creeper*
留鸟。尾上因具明显横斑而易与其他旋木雀相区别。见于中高海拔针阔混交林和针叶林。摄影／胡万新
A resident, the bar-tailed bird has apparent marks on its plumage different from other treecreepers. They are found in mid-high altitude needleleaf forests and mixed needleleaf- broadleaf forests. Photography by Hu Wanxin

涉禽

Waders

涉禽习惯于在浅水或岸边栖息生活，嘴、脚和颈部较长，有涉水捕食的生活习性。长青保护区有涉禽 36 种，常见于区内河流溪谷和周边塘库湿地，大多为候鸟迁徙经停，部分在此繁殖。涉禽中有国家一级保护动物 1 种，二级重点保护动物 2 种。

Waders (Shorebirds) are adapted to living by the shallows or coastal habitats. They have noticeably longer bills, legs and necks, and are accustomed to wading on the shores when feeding. The Changqing Nature Reserve is home to 36 species of waders which are found in rivers, ravines, and surrounding ponds or wetlands. Most species are passage migrants which land in the reserve for rest while some mate and reproduce during this time. Among them are one Class I state protected species and two Class II state protected species.

1. 白琵鹭 White Spoonbill *Platalea alba*
候鸟，琵琶形的灰色嘴特征明显。冬季见于汉江河谷地带，区内偶见。摄影／赵纳勋
White spoonbills are migratory birds with a unique, spoon-shaped bill grey in color. They are seen in the winter along the Han River shores and occasionally found in the nature reserve. Photography by Zhao Naxun

2. 白胸苦恶鸟 White-breasted Waterhen *Amaurornis phoenicurus*
夏候鸟。初夏时节常见于区内河谷湿地及稻田。摄影／赵纳勋
Being a summer migrant, white-breasted waterhens are found in the nature reserve in early summers, near ravines, wetlands or paddy fields. Photography by Zhao Naxun

3. 白鹭 Little Egret *Egretta garzetta*
中等体型（60厘米）鹭鸟，区内低海拔河谷地带常见。摄影／赵纳勋
Medium-sized (60 centimeters) little egrets are often sighted within the nature reserve near lower altitude ravines. Photography by Zhao Naxun

4. 中白鹭 Intermediate Egret *Mesophoyx intermedia*
体型大小在白鹭与大白鹭之间，喜稻田、塘库及沼泽地。摄影／赵纳勋
Between the size of great and small egret, the intermediate egret enjoys paddy fields, reservoirs and swamps. Photography by Zhao Naxun

1	2
3	4

秦岭长青
野生生命的庇护所
CHANGQING, QINLING
A Natural Shelter to Wildlife

1	2		6	7
3	4			
5				

1. 扇尾沙锥 Common Snipe *Gallinago gallinago*
候鸟。栖息于河岸、湖泊边、沼泽及水田地，多在晨昏和夜间活动，草丛中筑巢。摄影／赵纳勋
Common snipes are migrant waders that inhabit river banks, lakesides, swamps and paddy fields. They are active during dawn and dusk, and build their nests in the grass. Photography by Zhao Naxun

2. 针尾沙锥 Pintail Snipe *Gallinago stenura*
候鸟。常光顾区内及周边稻田、林中的沼泽和潮湿洼地。摄影／赵纳勋
Pintail snipes are migrant waders and frequent visitors to the nature reserve grounds and the surrounding paddy fields, swamps and marshes. Photography by Zhao Naxun

3，4. 夜鹭 Black-crowned Night-Heron *Nycticorax nycticorax*
头大体壮、色彩分明。本地分布广、数量多，区内常见于河谷地带。摄影／赵纳勋
Heavily built with a large head and contrasting plumage, this heron has a large population widely distributed in the area. They are found along the ravines. Photography by Zhao Naxun

5. 泽鹬 Marsh Sandpiper *Tringa stagnatilis*
过境鸟，春季偶见于华阳、茅坪低海拔沼泽和稻田。摄影／赵纳勋
Passage migrant bird, occasionally seen at low altitude marsh and rice field of Huayang and Maoping. Photography by Zhao Naxun

6. 绿鹭 Striated Heron *Butorides striata*
繁殖于本地，常见于区内低海拔河流溪谷。摄影／赵纳勋
Breeding in the area, striated heros are found in the nature reserve's ravines usually located at lower altitudes. Photography by Zhao Naxun

7. 骨顶鸡 Common Coot *Fulica atra*
候鸟。保护区内杨家沟、吊坝河和茅坪河春季常见，也曾发现于海拔2300米的针叶林带并被拍摄，为迁徙经停。摄影／胡万新
Common coots are passage migratory waders often sighted in Yangjiagou, the Diaoba River and the Maoping River during spring time. Their appearance have also been sighted and captured on camera in needleleaf forests 2,300 meters above seal level. Photography by Hu Wanxin

1. 苍鹭 Grey Heron *Ardea cinerea*
深沉的呱呱声及似鹅的叫声传播较远，常凝神站立于浅水区捕食。区内及周边四季常见。摄影／赵纳勋
The grey heron has a deep voice that produces a croaking sound that travels a long distance. It is often seen standing still in shallow waters waiting for aquatic preys. They can be found in the nature reserve throughout the year. Photography by Zhao Naxun

2. 白腰草鹬 Green Sandpiper *Tringa ochropus*
飞行时清晰可见黑色的下翼、白色的腰部。冬春季节常见于河流、溪谷以及沿岸稻田湿地。摄影／张永文
When in flight, the green sandpiper displays its distinct black wingspan and white abdomen. It is found along the shores of rivers, ravines, paddy fields and wetlands. Photography by Zhang Yongwen

3. 池鹭 Chinese Pond-Heron *Ardeola bacchus*
夏候鸟。常见活动于区内及周边溪流河谷、沼泽地和稻田。摄影／赵纳勋
A summer migrant, the Chinese pond-heron is often seen active in the nature reserve, around streams ravines, swamps and paddy fields. Photography by Zhao Naxun

4. 黑水鸡 Common Moorhen *Gallinula chloropus*
候鸟。春暖花开时光临，活动于华阳吊坝河、杨家沟有宽阔水面河谷。摄影／赵纳勋
These migratory birds visit the nature reserve in the spring and summer. They are often seen in wider streams and ravines near the Diaoba River and Yangjiagou in Huayang. Photography by Zhao Naxun

5. 大麻鳽 Great Bittern *Botaurus stellaris*
候鸟。为体大的金褐色及黑色鳽，偶见于区内。摄影／赵纳勋
Great bitterns are large migratory waders with a buffy-brown plumage covered with dark streaks and bars. They are frequently sighted in the nature reserve. Photography by Zhao Naxun

6. 凤头麦鸡 Northern Lapwing *Vanellus vanellus*
候鸟。凤头反翻，上体呈黑色金属光泽。分布于低海拔河流及塘库沿岸。摄影／赵纳勋
A migrant, the northern lapwing has a back-pointing crest and covered black plumage with an iridescent metallic sheen. These birds are distributed along the banks of low altitude rivers and reservoirs. Photography by Zhao Naxun

7. 大沙锥 Swinhoe's Snipe *Gallinago megala*
冬候鸟。嘴长似锥，区内及周边河流沿岸沼泽、稻田常见。摄影／赵纳勋
A winter migrant with a tapered bill, this shorebird can be found in the nature reserve, along the shores of rivers, swamps, paddy fields. Photography by Zhao Naxun

8. 黑翅长脚鹬 Black-winged Stilt *Himantopus himantopu*
冬候鸟。修长高挑，姿态优雅，于浅水沼泽地觅食。摄影／赵纳勋
Black-winged stilts migrate in the winter. They are slender and tall in form, and elegant in motion. They feed in shallows and swamps. Photography by Zhao Naxun

9. 反嘴鹬 Pied Avocet *Recurvirostra avosetta*
过境鸟。嘴细长而上翘。偶见于区内河谷地带。摄影／赵纳勋
Pied avocets are passage migrants with long bills curving upwards. They are occasionally sighted in the ravines within the nature reserve. Photography by Zhao Naxun

野生生命的庇护所
CHANGQING, QINLING
A Natural Shelter to Wildlife

野生生命的庇护所
CHANGQING, QINLING
A Natural Shelter to Wildlife

1. 矶鹬 Common Sandpiper *Actitis hypoleucos*
旅鸟。见于保护区中低海拔稻田和溪流岸边，行走时头不停地点动。摄影／张永文
The common sandpiper is a passage migrant shorebird, sometimes sighted in the nature reserve's mid-low altitude paddy fields and streams. They walk with a distinctive teeter, bobbing up and down constantly. Photography by Zhang Yongwen

2. 普通秧鸡 Water Rail *Rallus aquaticus*
候鸟。拍摄于华阳红石窑河，低海拔河流溪谷偶见。摄影／孙承骞
This migrant wader was captured on camera near Huayang's Hongshiyao River. Water rails are mostly seen near ravines at lower altitudes. Photography by Sun Chengqian

3. 林鹬 Wood Sandpiper *Tringa glareola*
候鸟。偶见于河谷地带的沼泽地和稻田。摄影／向定乾
A migratory bird, wood sandpipers are occasionally sighted along ravines, swamps and by paddy fields. Photography by Xiang Dingqian

4. 牛背鹭 Cattle Egret *Bubulcus ibis*
本地优势种，区内外广泛分布，极常见。摄影／赵纳勋
A dominant species in the area, cattle egrets are widely distributed within the nature reserve and beyond. They are frequently seen. Photography by Zhao Naxun

5. 灰头麦鸡 Grey-headed Lapwing *Vanellus cinereus*
候鸟。保护区低海拔河谷及周边社区常见，集小群活动于开阔沼泽、水田和耕地。摄影／赵纳勋
Grey-headed lapwings are migrant waders often seen in the lower altitude ravines and near communities. They appear in small paddlings in open swamps, paddy fields, and crop fields. Photography by Zhao Naxun

6. 金眶鸻 Little Ringed Plover *Charadrius dubius*
候鸟。黄色眼圈及黑褐色金胸带明显，夏季常见于河流沿岸。摄影／张永文
The little ringed plover is a migrant wader with yellow rings around its eyes and golden brown stripes on its chest. They are found by the rivers along the shore in the summer. Photography by Zhang Yongwen

7. 鹮嘴鹬 Ibisbill *Ibidorhyncha struthersii*
留鸟。分布于华阳、茅坪低海拔河谷地带，数量较少。摄影／赵纳勋
The ibisbills are non-migratory birds found mainly in low altitude ravines near Huayang and Maoping. Only a small population live in the area. Photography by Zhao Naxun

1	2		
3	4	6	7
5			

摄影/赵纳勋 Photography by Zhao Naxun

游禽

Waterbirds

游禽又称水禽，善于在水中游泳或潜水捕食。长青保护区已记录游禽 32 种，大多是自汉江沿酉水河谷游荡于此，部分为季节性迁徙至此。游禽在长青保护区主要分布于低海拔河谷地带。其中属于国家一级和二级保护动物各 1 种。

Waterbirds, also known as aquatic birds, are adapted to living and feeding in water. The Changqing Nature Reserve has recorded 32 species of waterbirds. Most of these bird populations have travelled downstream from the Han River through the Youshui River to local watercourses. Others are visiting Changqing due to seasonal migrations. Waterbirds are distributed mainly in the streams in lower altitudes of the reserve streams. Among their number include one Class Ⅰ state and one Class Ⅱ state protected species.

1. 普通鸬鹚 Great Cormorant *Phalacrocorax carbo*
又称鱼鹰、水老鸭，单独或结群在河中潜水捕鱼，冬春季见于区内低海拔河流溪谷。摄影／赵纳勋
Also known as great black cormorant, these birds dive deep into the water to catch fish. They hunt alone sometimes or in groups other times. They are often sighted in lower-altitude rivers or ravines within the reserve. Photography by Zhao Naxun

2. 赤麻鸭 Ruddy Shelduck *Tadorna ferruginea*
候鸟，酉水河下游及汉江极常见，区内偶见，以各种谷物、昆虫、甲壳动物、蛙、虾、水生植物为食。摄影／赵纳勋
The ruddy shelducks are migratory birds often seen in the lower reaches of the Youshui River or along the Han River. They are occasionally seen in the nature reserve and their food source include crops, insects, crustaceans, frogs, shrimp and aquatic plants. Photography by Zhao Naxun

3. 凤头䴙䴘 Great Crested Grebe *Podiceps cristatus*
候鸟，中型游禽，两束黑色长形冠羽特征明显。主要以鱼类和水生无脊椎动物为食。摄影／赵纳勋
The great crested grebe is a medium-sized migratory waterbird with two distinct strips of feathery black crests on its forehead. It lives on fish and aquatic invertebrates as its primary food source. Photography by Zhao Naxun

4. 红嘴鸥 Common Black-headed Gull *Chroicocephalus ridibundus*
候鸟，灰白色，嘴及脚红色，繁殖羽具有深棕色的头罩，偶见于低海拔河流、塘库。摄影／赵纳勋
The common black-headed gull is a migrant bird with a pale grey body, red bill and red feet. Its breeding plumage exhibits a dark brown hood. They are occasionally sighted in lower altitude rivers and reservoirs. Photography by Zhao Naxun

	2
1	3
	4

野生生命的庇护所
CHANGQING, QINLING
A Natural Shelter to Wildlife

1. 小䴙䴘 Little Grebe *Tachybaptus ruficollis*
留鸟，低海拔河流、塘库常见，善潜水以鱼类和水生昆虫为食。摄影／赵纳勋
Little grebes are small non-migratory birds often sighted in lower altitude rivers and reservoirs. They are keen divers whose diet include fish and aquatic insects. Photography by Zhao Naxun

2. 绿头鸭 Mallard *Anas platyrhynchos*
冬候鸟，雄鸟头顶带绿色光泽，白色颈环明显，又叫大野鸭。偶见于区内低海拔河流湖泊。摄影／胡万新
Mallards is winter migrant. The male bird has a glossy green crown and a white neck ring is prominent, also called wild ducks. They are occasionally sighted in rivers and lakes located in lower altitudes in the nature reserve. Photography by Hu Wanxin

3. 普通秋沙鸭 Common Merganser *Mergus merganser*
候鸟，冬春季节常见于区内酉水河及其主要支流，集小群活动，以鱼为食。摄影／赵纳勋
Migratory, these birds are primarily found during spring and winter in the nature reserve's Youshui River and its tributaries. They travel in small paddlings or flocks and feed on fish. Photography by Zhao Naxun

4. 鸳鸯 Mandarin Duck *Aix galericulata*
候鸟，国家二级保护动物。冬春季节常见于保护区内较低海拔的河流溪谷。摄影／赵纳勋
Mandarin ducks are migratory birds primarily found during spring and winter in the nature reserve's lower altitude rivers and ravines. Photography by Zhao Naxun

5. 普通燕鸥 Common Tern *Sterna hirundo*
候鸟，体型较小（35厘米）头顶黑色的燕鸥。夏初偶见于保护区低海拔水面宽阔的河流沿岸。摄影／赵纳勋
A small migratory bird (35 centimeters), the common tern has a black crown. They are occasionally sighted in early summers along the shores of wider rivers located in low altitudes. Photography by Zhao Naxun

6. 翘鼻麻鸭 Common Shelduck *Tadorna tadorna*
嘴上翘赤红色，基部生有一个突出的红色皮质瘤，颜色艳丽。偶见于区内低海拔河谷地区，冬候鸟。摄影／赵纳勋
With a curved red bill bearing a prominent knob at the forehead, this water fowl has vibrant colors on its body and are sometimes found active in the nature reserve's lower ravines. Photography by Zhao Naxun

7. 绿翅鸭 Common Teal *Anas crecca*
冬候鸟，善集群活动，羽色较艳丽，飞行时绿色翼镜明显。常光顾区内杨家沟等低海拔河流与池塘。摄影／赵纳勋
These winter migrants are gregarious birds with bright plumage. When in flight, these teals display vivid green speculum feathers. They often visit the low altitude rivers and ponds near Yangjiagou within the nature reserve. Photography by Zhao Naxun

8. 中华秋沙鸭 Scaly-sided Merganser *Mergus squamatus*
候鸟，国家一级保护动物，长长的冠羽和树洞筑巢的习性有别于鸭科其他种。偶见于低海拔河谷。摄影／赵纳勋
Class I state protected species. These migratory waterbirds have a long feathered crest and live in tree hollows, which distinguishes themselves from other ducks. They are occasionally seen in lower altitude ravines or river valleys. Photography by Zhao Naxun

碧凤蝶 *Papilio bianor bianor* Cramer
摄影／赵纳勋
Photography by Zhao Naxun

野生生命的庇护所
CHANGQING, QINLING
A Natural Shelter to Wildlife

昆虫类
Insects

昆虫是动物界中种类最多的一个纲，已知长青保护区内分布有昆虫12目95科636种，其中蝶类164种。保护区内蝶类主要分布于海拔800～2500米的地带，因其灵动飘逸的姿态、绚丽多姿的色彩，成为保护区内会飞的花朵，在碾子湾、破岔峪和柏杨坪一带分布较为集中，是该区域丰富生物多样性的鲜活体现，具有很高的观赏价值。

Insects are the largest class of animals in the animal kingdom. In the Changqing Nature Reserve, 636 species in 95 families and 12 orders have been discovered. Among them are 164 butterfly species, mostly distributed in areas from 800 to 2,500 meters above sea level. The lively and graceful movements of the butterflies in their brilliant colors make them appear like hovering flowers in the nature reserve grounds. Their great populations in and around Nianziwan, Pochayu and Baiyangping point to great biodiversity in these areas.

秦岭长青 野生生命的庇护所
CHANGQING, QINLING
A Natural Shelter to Wildlife

1. 白斑迷蛱蝶 *Mimathyma schrenckii* Ménétriès
   摄影／胡万新 Photography by Hu Wanxin

2. 斑缘豆粉蝶 *Colias erate* Esper
   摄影／赵纳勋 Photography by Zhao Naxun

3. 冰清绢蝶 *Parnassius glacialis* Butler
   摄影／向定乾 Photography by Xiang Dingqian

4. 菜粉蝶 *Pieris rapae* Linnaeus
   摄影／赵纳勋 Photography by Zhao Naxun

5. 大红蛱蝶 *Vanessa indica* Herbst
   摄影／赵纳勋 Photography by Zhao Naxun

6. 大翅绢粉蝶 *Aporia largeteaui* Oberthür
   摄影／杜小健 Photography by Du Xiaojian

7. 巴黎翠凤蝶 *Papilio paris* Linnaeus
   摄影／时鉴 Photography by Shi Jian

8. 灿福蛱蝶 *Fabriciana adippe* Denis et Schiffermuller
   摄影／张永文 Photography by Zhang Yongwen

9. 翠蓝眼蛱蝶 *Junonia orithya* Linnaeus
   摄影／赵纳勋 Photography by Zhao Naxun

10. 橙黄豆粉蝶 *Colias fieldii* Ménétriés
    摄影／胡万新 Photography by Hu Wanxin

秦岭长青
野生生命的庇护所
CHANGQING, QINLING
A Natural Shelter to Wildlife

1	2		7		
3	4		8	9	10
5	6				

1. 黛眼蝶 *Lethe dura* Marshall
摄影／向定乾 Photography by Xiang Dingqian

2. 大卫绢蛱蝶 *Calinaga davidis* Oberthür
摄影／胡万新 Photography by Hu Wanxin

3. 稻眉眼蝶 *Mycalesis gotama* Moore
摄影／赵纳勋 Photography by Zhao Naxun

4. 斐豹蛱蝶（雄）*Aegyreus hyperbius* Linnaeus
摄影／胡万新 Photography by Hu Wanxin

5. 黑纹粉蝶 *Pieris melete* Ménétriés
摄影／张永文 Photography by Zhang Yongwen

6. 黑带黛眼蝶 *Lethe nigrifascia* Leech
摄影／赵纳勋 Photography by Zhao Naxun

7. 柑橘凤蝶 *Papilio xuthus* Linnaeus
摄影／赵纳勋 Photography by Zhao Naxun

8. 大紫蛱蝶 *Sasakia charonda* Hewison
摄影／胡万新 Photography by Hu Wanxin

9. 东方菜粉蝶 *Pieris canidia* Sparrman
摄影／赵纳勋 Photography by Zhao Naxun

10. 黑弄蝶 *Daimio tethys* Ménétriés
摄影／赵纳勋 Photography by Zhao Naxun

野生生命的庇护所
CHANGQING, QINLING
A Natural Shelter to Wildlife

1. 箭纹粉眼蝶 *Callarge sagitta* Leech
摄影／赵纳勋 Photography by Zhao Naxun

2. 红基美凤蝶（雄）*Papilio alcmenor* Felder et Felder
摄影／刘伟 Photography by Liu Wei

3. 黄钩蛱蝶 *Polygonia c-aureum* Linnaeus
摄影／向定乾 Photography by Xiang Dingqian

4. 箭环蝶 *Stichophthalma howqua* Westwood
摄影／赵纳勋 Photography by Hu Zhao Naxun

5. 黄环蛱蝶 *Neptis themis* Leech
摄影／向定乾 Photography by Xiang Dingqian

6. 金凤蝶 *Papilio machaon* Linnaeus
摄影／胡万新 Photography by Hu Wanxin

7. 黄帅蛱蝶 *Sephisa princeps* Fixsen
摄影／向定乾 Photography by Xiang Dingqian

8. 金裳凤蝶（雄）*Troides aeacus* Felder et Felder
摄影／赵纳勋 Photography by Zhao Naxun

9. 金裳凤蝶（雌）*Troides aeacus* Felder et Felder
摄影／赵纳勋 Photography by Zhao Naxun

1. 美线蛱蝶 *Limenitis misuji* Sugiyama
摄影／向定乾 Photography by Xiang Dingqian

2. 酪色绢粉蝶 *Aporia potanini* Alphéraky
摄影／向定乾 Photography by Xiang Dingqian

3. 琉璃蛱蝶 *Kaniska canace* Linnaeus
摄影／向定乾 Photography by Xiang Dingqian

4. 绿豹蛱蝶 *Argynnis paphia* Linnaeus
摄影／赵纳勋 Photography by Zhao Naxun

5. 宽尾凤蝶 *Agehana elwesi* Leech
摄影／向定乾 Photography by Xiang Dingqian

6. 宽边黄粉蝶 *Eurema hecabe* Linnaeus
摄影／赵纳勋 Photography by Zhao Naxun

7. 蓝灰蝶 *Everes argiades* Pallas
摄影／赵纳勋 Photography by Zhao Naxun

8，9. 蓝凤蝶 *Papilio protenor* Fruhstorfer
摄影／赵纳勋 Photography by Zhao Naxun

10. 灰姑娘绢粉蝶 *Aporia interconstata* Bang-haas
摄影／胡万新 Photography by Hu Wanxin

1	2	5	8
3	4	6	9
		7	10

野生生命的庇护所
CHANGQING, QINLING
A Natural Shelter to Wildlife

1. 朴喙蝶 *Libythea lepita* Moore
摄影／向定乾 Photography by Xiang Dingqian

2. 宁眼蝶 *Ninguta schrenkii* Menetries
摄影／张希明 Photography by Zhang Ximing

3. 摩来彩灰蝶 *Heliophorus moorei* Hewitson
摄影／赵纳勋 Photography by Zhao Naxun

4. 曲带闪蛱蝶 *Apatura laverna* Leech
摄影／向定乾 Photography by Xiang Dingqian

5. 秦岭绢粉蝶 *Aporia tsinglingica* Verity
摄影／赵纳勋 Photography by Zhao Naxun

6. 双色舟弄蝶 *Barca bicolor* Oberthür
摄影／赵纳勋 Photography by Zhao Naxun

7. 双星箭环蝶 *Stichophthalma neumogeni* Leech
摄影／向定乾 Photography by Xiang Dingqian

8. 麝凤蝶 *Byasa alcinous confuse* Klug
摄影／向定乾 Photography by Xiang Dingqian

9. 突缘麝凤蝶 *Byasa plutonius* Oberthür
摄影／赵纳勋 Photography by Zhao Naxun

秦岭长青 野生生命的庇护所
CHANGQING, QINLING
A Natural Shelter to Wildlife

```
1   3   8
    4
    5   9   10
    6
2   7   11  12
```

1. 乌克兰剑凤蝶 *Pazala tamerlana* Oberthür
摄影／胡万新 Photography by Hu Wanxin

2. 网眼蝶 *Rhaphicera dumicola* Oberthür
摄影／张希明 Photography by Zhang Ximing

3. 丫纹绢粉蝶 *Aporia delavayi* Oberthür
摄影／赵纳勋 Photography by Zhao Naxun

4. 雾驳灰蝶 *Bothrinia nebulosa* Lecch
摄影／赵纳勋 Photography by Zhao Naxun

5. 长纹电蛱蝶 *Dichorragia nesseus* Grose-Smith
摄影／胡万新 Photography by Hu Wanxin

6. 小红蛱蝶 *Vanessa cardui* Linnaeus
摄影／赵纳勋 Photography by Zhao Naxun

7. 重环蛱蝶 *Neptis alwina* Bremer
摄影／赵纳勋 Photography by Zhao Naxun

8. 窄斑翠凤蝶（雄）*Papilio arcturus* Westwood
摄影／胡万新 Photography by Hu Wanxin

9. 直纹稻弄蝶 *Parara guttata* Bremeret Grey
摄影／赵纳勋 Photography by Zhao Naxun

10. 圆翅钩粉蝶 *Gonepteryx amintha* Blanchard
摄影／赵纳勋 Photography by Zhao Naxun

11. 玉带凤蝶（雌）*Papilio polytes* Linnaeus
摄影／时鉴 Photography by Shi Jian

12. 中华黄葩蛱蝶 *Patsuia sinensis* Oberthür
摄影／赵纳勋 Photography by Zhao Naxun

独叶草 *Kingdonia uniflora* Balf. f. et W. W. Smith
摄影/赵纳勋 Photography by Zhao Naxun

五

长青的奇葩

The Wonders of Changqing

由于地处暖温带与亚热带的过渡区，过渡性植被特征十分明显。长达 20 多年的强度采伐和森林抚育以及保护区建立以来的栖息自然恢复过程中，区内的环境和植被发生了很大变化。表现为植被类型复杂，阔叶林面积大、分布广、种类多，中低海拔人工林多。特别是高海拔寒温性针叶林保存完整，林下分布大面积秦岭箭竹，是秦岭大熊猫非常优良的夏居地，具有很高的保护和研究价值。

Located in the transition area between the warm temperate zone and the subtropics, Changqing's vegetation has distinct characteristics. 20 years of intensive logging, cultivation and natural restoration after the establishment of the nature reserve have altered the environment and vegetation dramatically. The resulting manifestations include a more complex mix of vegetation types, increase in acreage, distribution and plant species of broadleaf forests, increased man-made forests at mid-low latitudes. Another significant feature is the well-preserved high altitude cold temperate needleleaf forests, which accommodate great distributions of *F.qinlingensis* (bamboo species) that makes these forests an ideal summer habitat for Qinling giant pandas, thus making these forests highly valued for both conservation and research.

摄影／赵纳勋
Photography by Zhao Naxun

植被垂直分布
The Vertical Distribution of Vegetation

1. 针阔混交林 Mixed coniferous
摄影／赵纳勋 Photography by Zhao Naxun

2. 针叶林 Coniferous forest
摄影／赵纳勋 Photography by Zhao Naxun

3. 巴山木竹林 Fargesii forest
摄影／赵纳勋 Photography by Zhao Naxun

4. 高山草甸 Alpine meadow
摄影／赵纳勋 Photography by Zhao Naxun

5. 阔叶林 Broad-leaved forest
摄影／赵纳勋 Photography by Zhao Naxun

1. 苦糖果 *Lonicera fragrantissima* subsp. *standishii* (Carr.) Hsu et H. J. Wang
摄影 / 赵纳勋 Photography by Zhao Naxun

2. 望春玉兰 *Magnolia biondii* Pamp.
摄影 / 赵纳勋 Photography by Zhao Naxun

3. 笔龙胆 *Gentiana zollingeri* Fawcett
摄影 / 董伟 Photography by Dong Wei

时间的花序
The Inflorescence of Time

独特的地理位置，多样的地貌类型，温暖湿润的气候，造就以长青保护区为代表的秦岭山地植物种类繁多，观赏植物异常丰富，一年四季均能欣赏到色泽斑斓、风姿绰约、香馨四溢的野生植物花卉。

Unique geographic location coupled with differing landscapes and mild, humid weather have enhanced the growth of diverse and bountiful exotic plants in Qinling, whose most typical ecological area is Changqing. All year round, visitors can enjoy rich medleys of colors, varying styles and lush scents of wild flowers .

一、二月
January & February

新年伊始，高山仍被积雪覆盖，而低海拔沉寂的万物已经开始萌动。望春玉兰、苦糖果、笔龙胆、春兰等已经含苞待放，预示着春天的脚步已经临近，生命开始复苏。

At the beginning of the new year, mountain tops are still capped with snow, dormant creatures closer to sea level are already awake in motion. Biond's magnolias, winter honeysuckles, *Gentiana zollingeri* and noble orchids are already budding, heralding the coming of spring and the rejuvenation of life.

三月

March

阳春三月,在春风的吹拂下,植物开始孕育花蕾、吐露绿色。白屈菜、白头翁、紫荆、铁筷子、野豌豆、山茱萸等植物的花朵已经悄悄绽放,开始装扮长青的山水。

In March, in the warm breeze of spring, while most wild plants are beginning to bud, heralding the advent of the green season, the greater celandines, Chinese pasque flowers, Chinese redbuds, Tibetan hellebores, bush vetches and cornelian cherries are already in bloom, adorning Changqing's mountains and rivers.

1. 秦岭木姜子 *Litsea tsinlingensis* Yang et P. H. Huang
摄影／张永文 Photography by Zhang Yongwen

2. 春兰 *Cymbidium goeringii* (Rchb. f.) Rchb. f.
摄影／杜小健 Photography by Du Xiaojian

3. 紫花地丁 *Viola philippica* Cav.
摄影／赵纳勋 Photography by Zhao Naxun

4. 荷青花 *Hylomecon japonica* (Thunb.) Prantl et Kundig
摄影／赵纳勋 Photography by Zhao Naxun

5. 白屈菜 *Chelidonium majus* Linn.
摄影／赵纳勋 Photography by Zhao Naxun

6. 大叶金腰 *Chrysosplenium macrophyllum* Oliv.
摄影／董伟 Photography by Dong Wei

7. 铁筷子 *Helleborus thibetanus* Franch.
摄影／赵纳勋 Photography by Zhao Naxun

四月

April

四月的长青已经春意盎然，风如酥，花似火。粉白欲滴的野桃花，随风摇曳的海棠花，还有那报春、黄堇、棣棠、杜鹃、映山红等竞相开放，将春天的气息撒满整座山林。

The vitality of April, inspirits Changqing transforming the mountains into a landscape of vernal beauty with gentle winds and blazing blooms. Wild peach trees are bursting with milky pink flowers. Chinese crabapple flowers are swaying in the breeze. Primroses, corydalis pallidas, Japanese roses, azaleas and evergreen azaleas are blooming with vigor, charging the forest with their fragrance.

	2
1	3
	4

1. 蕙兰 *Cymbidium faberi* Rolfe
摄影／赵纳勋 Photography by Zhao Naxun

2. 甘肃瑞香 *Daphne tangutica* Maxim.
摄影／赵纳勋 Photography by Zhao Naxun

3. 棣棠 *Kerria japonica* (Linn.) DC.
摄影／赵纳勋 Photography by Zhao Naxun

4. 齿萼报春 *Primula odontocalyx* (Franch.) Pax
摄影／赵纳勋 Photography by Zhao Naxun

1. 裂叶地黄 *Rehmannia piasezkii* Maxim.
摄影／赵纳勋 Photography by Zhao Naxun

2. 蒲儿根 *Sinosenecio oldhamianus* (Maxim.) B. Nord.
摄影／赵纳勋 Photography by Zhao Naxun

3. 中国旌节花 *Stachyurus chinensis* Franch.
摄影／董伟 Photography by Dong Wei

4. 领春木 *Euptelea pleiosperma* Hook. f. et Thomson
摄影／刘伟 Photography by Liu Wei

5. 太白杜鹃 *Rhododendron purdomii* Rehd. et Wils.
摄影／赵纳勋 Photography by Zhao Naxun

6. 四川杜鹃 *Rhododendron sutchuenense* Franch.
摄影／赵纳勋 Photography by Zhao Naxun

7. 紫斑牡丹 *Paeonia rockii* (S.G.Haw & Lauener) T.Hong & J.J.Li ex D.Y.Hong
摄影／赵纳勋 Photography by Zhao Naxun

8. 紫堇 *Corydalis edulis* Maxim.
摄影／胡万新 Photography by Hu Wanxin

9. 马桑 *Coriaria nepalensis* Wall.
摄影／董伟 Photography by Dong Wei

10. 野樱桃 *Cerasus pseudocerasus* (Lindl.) G. Don
摄影／赵纳勋 Photography by Zhao Naxun

11. 秀丽莓 *Rubus amabilis* (Focke)
摄影／董伟 Photography by Dong Wei

12. 三叶木通 *Akebia trifoliata* (Thunb.) Koidz.
摄影／赵纳勋 Photography by Zhao Naxun

13. 酢浆草 *Oxalis corniculata* Linn.
摄影／董伟 Photography by Dong Wei

14. 紫罗兰报春 *Primula purdomii* Craib
摄影／赵纳勋 Photography by Zhao Naxun

1	3		6	7	8
4			9	10	11
2	5		12	13	14

五月

May

五月的风从春天的阳光中走来，把长青的树叶儿吹绿，将长青的花儿吹红。依次开放的花朵，红的像火，粉的像霞，白的像雪。长青的山林如诗如画，充满着勃勃生机。

The wind of May mingles with the sun rays, dying the leaves green and flowers red. The flowers bloom in colors, ranging from fiery red, to pink and to snow white. During this month, Changqing bursts with life, and its landscape is like a poetic painting, full of vitality and vigour.

1	2	5	6
3	4	7	8
		9	10

野生生命的庇护所
CHANGQING, QINLING
A Natural Shelter to Wildlife

1. 杠柳 *Periploca sepium* Bunge
摄影／向定乾 Photography by Xiang Dingqian

2. 华北楼斗菜 *Aquilegia yabeana* Kitagawa
摄影／赵纳勋 Photography by Zhao Naxun

3. 厚朴 *Houpoëa officinalis* (Rehder et E. H. Wilson) N. H. Xia et C. Y.
摄影／赵纳勋 Photography by Zhao Naxun

4. 黄素馨 *J. giraldii* Diels
摄影／赵纳勋 Photography by Zhao Naxun

5. 列当 *Orobanche coerulescens* Steph.
摄影／董伟 Photography by Dong Wei

6. 盘叶忍冬 *Lonicera tragophylla* Hemsl.
摄影／赵纳勋 Photography by Zhao Naxun

7. 山竹花 *Disporum cantoniense* (Lour.) Merr.
摄影／赵纳勋 Photography by Zhao Naxun

8. 四照花 *Cornus kousa* subsp. *chinensis* (Osborn) Q. Y. Xiang
摄影／赵纳勋 Photography by Zhao Naxun

9. 流苏虾脊兰 *Calanthe alpina* Hook. f. ex Lindl.
摄影／赵纳勋 Photography by Zhao Naxun

10 辽东丁香 *Syringa wolfii* Schneid.
摄影／赵纳勋 Photography by Zhao Naxun

野生生命的庇护所
CHANGQING, QINLING
A Natural Shelter to Wildlife

1. 三棱虾脊兰 *Calanthe tricarinata* Lindl.
摄影／胡万新 Photography by Hu Wanxin

2. 鸢尾 *Iris tectorum* Maxim.
摄影／张永文 Photography by Zhang Yongwen

3. 延龄草 *Trillium tschonoskii* Maxim.
摄影／赵纳勋 Photography by Zhao Naxun

4. 玉竹 *Polygonatum odoratum* (Mill.) Druce
摄影／赵纳勋 Photography by Zhao Naxun

5. 透骨消 *Glechoma longituba* (Nakai) Kupr.
摄影／赵纳勋 Photography by Zhao Naxun

6. 云南大百合 *Cardiocrinum giganteum* var. *yunnanense* (Leichtlin ex Elwes) Stearn
摄影／赵纳勋 Photography by Zhao Naxun

7. 朱兰状独蒜兰 *Pleione bulbocodioides* (Franch.) Rolfe
摄影／赵纳勋 Photography by Zhao Naxun

8. 中华绣线梅 *Neillia sinensis* Oliv.
摄影／赵纳勋 Photography by Zhao Naxun

9. 梓木草 *Lithospermum zollingeri* DC.
摄影／董伟 Photography by Dong Wei

10. 藓生马先蒿 *Pedicularis muscicola* Maxim.
摄影／赵纳勋 Photography by Zhao Naxun

1. 陕西报春 *Primula handeliana* W.W.Sm.et Forrest
摄影／向定乾 Photography by Xiang Dingqian

2. 舌唇兰 *Platanthera japonica* (Thunb. ex A. Murray) Lindl.
摄影／赵纳勋 Photography by Zhao Naxun

3. 长柄八仙花 *Hydrangea longipes* Franch.
摄影／赵纳勋 Photography by Zhao Naxun

4. 扇叶杓兰 *Cypripedium japonicum* Thunb.
摄影／胡万新 Photography by Hu Wanxin

5. 宝铎草 *Disporum sessile* D. Don
摄影／赵纳勋 Photography by Zhao Naxun

6. 南山藤 *Dregea volubilis* (Linn. f.) Benth. ex Hook. f.
摄影／赵纳勋 Photography by Zhao Naxun

六月

June

六月，一个万物鼎盛的月份，长青的森林里开满了各色各样的鲜花，灿烂得像撒满了宝石，铺上了锦缎，放眼一片璀璨。各种花卉争奇斗艳，漫山遍野，多彩绚烂。

June, a month of vibrancy for all life forms in the wild. The forests of Changqing are blooming with flowers of every color, scattered in the landscape and glittering like glorious gems that fill the sight. Covering the expanse of the slopes, plants of every kind seem to compete for attention by radiating a full scale of hues and tones.

野生生命的庇护所
CHANGQING, QINLING
A Natural Shelter to Wildlife

1. 大瓣铁线莲 *Clematis macropetala* Ledeb.
摄影／赵纳勋 Photography by Zhao Naxun

2. 单花无柱兰 *Amitostigma monanthum* (Finet) Schltr.
摄影／赵纳勋 Photography by Zhao Naxun

3. 点地梅 *Androsace umbellata* (Lour.) Merr.
摄影／赵纳勋 Photography by Zhao Naxun

1. 峨眉蔷薇 *Rosa omeiensis* Rolfe
摄影／赵纳勋 Photography by Zhao Naxun

2. 狗枣猕猴桃 *Actinidia kolomikta* (Maxim. et Rupr.) Maxim.
摄影／董伟 Photography by Dong Wei

3. 黄毛忍冬 *Lonicera acuminata* Wall.
摄影／胡万新 Photography by Hu Wanxin

4. 岷山银莲花 *Anemone rockii* Ulbr.
摄影／赵纳勋 Photography by Zhao Naxun

5. 头花杜鹃 *Rhododendron capitatum* Maxim.
摄影／董伟 Photography by Dong Wei

6. 毛杓兰 *Cypripedium franchetii* E. H. Wilson
摄影／任毅 Photography by Ren Yi

7. 黄花白芨 *Bletilla ochracea* Schltr.
摄影／赵纳勋 Photography by Zhao Naxun

8. 多花红升麻 *Astilbe rivularis* var. *myriantha* (Diels) J. T. Pan
摄影／赵纳勋 Photography by Zhao Naxun

9. 黑蕊猕猴桃 *Actinidia melanandra* Franch.
摄影／杜小健 Photography by Du Xiaojian

1	3	4	5
2		6	7
		8	9

1. 紫珠 *Callicarpa bodinieri* Lévl.
摄影／董伟 Photography by Dong Wei

2. 五脉绿绒蒿 *Meconopsis quintuplinervia* Regel
摄影／赵纳勋 Photography by Zhao Naxun

3. 无距耧斗菜 *Aquilegia ecalcarata* Maxim.
摄影／赵纳勋 Photography by Zhao Naxun

4. 圆叶鹿蹄草 *Pyrola rotundifolia* Linn.
摄影／胡万新 Photography by Hu Wanxin

5. 川赤芍 *Paeonia veitchii* Lynch.
摄影／赵纳勋 Photography by Zhao Naxun

6. 中华秋海棠 *Begonia grandis* subsp. *sinensis* (A. DC.) Irmsch.
摄影／赵纳勋 Photography by Zhao Naxun

7. 银露梅 *Potentilla glabra* Lodd.
摄影／赵纳勋 Photography by Zhao Naxun

七月

July

七月的长青花香袭人，植物的花期盛会随着海拔的升高逐渐上移，高山草甸也已成为花的海洋。色彩斑斓的花卉点缀着长青的角角落落，形成了一个绿的海洋、花的世界。

July in Changqing is marked by the fragrance from the flowers blooming in succession at different altitudes. And the alpine meadows also have transformed into an ocean of flowers. Like a green sea afloat with petals, the colorful ripples of flowers expand, infiltrating every corner of the mountains.

1. 卷丹 *Lilium tigrinum* Ker Gawl.
摄影／赵纳勋 Photography by Zhao Naxun

2. 瞿麦 *Dianthus superbus* Linn.
摄影／赵纳勋 Photography by Zhao Naxun

3. 拟缺香茶菜 *Isodon excisoides* (Y.Z. Sun ex C. H. Hu) H. Hara
摄影／杜小健 Photography by Du Xiaojian

4. 百合 *Lilium brownii* var. *viridulum* Baker
摄影／赵纳勋 Photography by Zhao Naxun

5. 大花糙苏 *Phlomis megalantha* Diels
摄影／赵纳勋 Photography by Zhao Naxun

6. 广布红门兰 *Orchis chusua* D. Don
摄影／董伟 Photography by Dong Wei

7. 弧距虾脊兰 *Calanthe arcuata* Rolfe
摄影／董伟 Photography by Dong Wei

1. 金丝桃 *Hypericum monogynum* Linn.
摄影／董伟 Photography by Dong Wei

2. 绶草 *Spiranthes sinensis* (Pers.) Ames
摄影／胡万新 Photography by Hu Wanxin

3. 水杨梅 *Geum aleppicum* Jacq.
摄影／董伟 Photography by Dong Wei

4. 太白山报春 *Primula giraldiana* Pax
摄影／赵纳勋 Photography by Zhao Naxun

5. 油点草 *Tricyrtis macropoda* Miq.
摄影／胡万新 Photography by Hu Wanxin

6. 珠根老鹳草 *Geranium pylzowianum* Maxim.
摄影／赵纳勋 Photography by Zhao Naxun

7. 紫萼女娄菜 *Melandrium tatarinowii* Regel
摄影／胡万新 Photography by Hu Wanxin

8. 商陆 *Phytolacca acinosa* Roxb.
摄影／赵纳勋 Photography by Zhao Naxun

八月

August

  盛夏的长青，葱茏、秀丽、多姿。蓝白相间的太白龙胆花，粉红色的马先蒿，紫色的松潘乌头花，以及剪秋萝、射干等鲜红艳丽的花朵，为苍翠欲滴的山林带来了一片繁华的景象。

  Mid-summer in Changqing is an opulent scene, picturesque and manifold. Blue and white Japanese gentiana flowers, pink lousewort flowers, purple aconitium sungpanenses, along with lychnis cognatas and leopard flowers, add to the crisp green forest a sense of prosperity.

1. 剪秋萝 *Lychnis cognata* Maxim.
摄影／赵纳勋 Photography by Zhao Naxun

2. 扭盔马先蒿 *Pedicularis davidii* Franch.
摄影／赵纳勋 Photography by Zhao Naxun

3. 川续断 *Dipsacus asperoides* Hoffmanns. & Link
摄影／胡万新 Photography by Hu Wanxin

4. 忽地笑 *Lycoris aurea* (L'Hér.) Herb.
摄影／赵纳勋 Photography by Zhao Naxun

1. 太白龙胆 *Gentiana apiata* N. E. Brown
摄影／董伟 Photography by Dong Wei

2. 松潘乌头 *Aconitum sungpanense* Hand.-Mazz.
摄影／张永文 Photography by Zhang Yongwen

3. 星叶草 *Circaeaster agrestis* Maxim.
摄影／董伟 Photography by Dong Wei

4 苍耳七 *Parnassia wightiana* Wall. ex Wight et Arn.
摄影／赵纳勋 Photography by Zhao Naxun

5. 射干 *Belamcanda chinensis* (L.) DC.
摄影／张永文 Photography by Zhang Yongwen

6. 轮叶马先蒿 *Pedicularis verticillata* Linn.
摄影／赵纳勋 Photography by Zhao Naxun

| 1 | 2 | 5 |
| 3 | 4 | 6 |

九月
September

　　九月的长青依然草木茂盛、百花争艳。晶莹剔透的水晶兰，鲜红的石蒜花，淡黄色的水金凤花正在开放，快要成熟的果实已挂满枝头，收获的季节即将来临。

　　When September arrives, Changqing's greenery and blooms are still thriving. Amidst the crystalline blooms of Indian pipes, bright red hues of red spider lilies and yellow touch-me-not balsams, a full array of fruits have already come out dangling on branches, announcing the approaching harvest.

1. 水金凤 *Impatiens noli-tangere* Linn.
摄影／赵纳勋 Photography by Zhao Naxun

2. 水晶兰 *Monotropa uniflora* Linn.
摄影／杜小健 Photography by Du Xiaojian

3. 花锚 *Halenia corniculata* (L.) Cornaz
摄影／胡万新 Photography by Hu Wanxin

4. 石蒜 *Lycoris radiata* (L'Hér.) Herb.
摄影／杜小健 Photography by Du Xiaojian

十月

October

金秋十月，野果飘香，山野在不觉间已变成一片斑斓。在这五彩缤纷的世界里，绽放的野菊花、乌头花、大火草花与丰收的果实一起为灿烂辉煌的金秋增色添彩。

During the golden month of October, wild fruits spill fragrance all over the forests. In an instant, the wilderness has turned into clusters of brilliant autumn colors. In this colorful world, chrysanthemums, Chinese aconites and *Anemone grapeleafs* bloom along the harvests of wild fruits, adding to the radiant golden hues of the scenery.

1. 野鸦椿 *Euscaphis japonica* (Thunb.) Kanitz
摄影／赵纳勋 Photography by Zhao Naxun

2. 大火草 *Anemone tomentosa* (Maxim.) Pei
摄影／赵纳勋 Photography by Zhao Naxun

3. 红豆杉 *Taxus wallichiana* var. *chinensis* (Pilg.) Florin
摄影／贺明锐 Photography by He Mingrui

4. 野菊花 *Chrysanthemum indicum* Thunb.
摄影／赵纳勋 Photography by Zhao Naxun

| 1 | 2 |
| 3 | 4 |

十一、十二月

November & December

短暂的金秋过后，长青的山林很快就迎来了冬天，各种动植物开始经受严冬的考验。当十二月冰雪覆盖的时候，大多数植物已进入休眠期，常绿植物却在风雪之中顽强地彰显着长青的生命活力。

Following the transcience of autumn, winter is now arriving into the mountains of Changqing. Animals and plants of every kind must now endure the challenge of the harsh winter. When December finally comes with its thick coat of ice and snow, most plants have entered into dormancy, with the exception of ever-green plants, which continue to persevere against the wind and snow, revealing the extraordinary vitality of life in Changqing.

野生生命的庇护所
CHANGQING, QINLING
A Natural Shelter to Wildlife

1. 冬日晨光 Morning sunshine in Winter
摄影／张永文 Photography by Zhang Yongwen

2. 风雪中的巴山冷杉 Outstanding fir in the snow and wind
摄影／胡万新 Photography by Hu Wanxin

3. 翠绿的巴山木竹 Emerald green bamboo
摄影／向定乾 Photography by Xiang Dingqian

古道仙境 Ancient heavenly path
摄影／赵纳勋 Photography by Zhao Naxun

野生生命的庇护所
CHANGQING, QINLING
A Natural Shelter to Wildlife

六
人与自然和谐的家园

Human and Nature in Harmony

长青不仅是秦岭大熊猫的野生家园,众多野生生命的自然庇护所,更是人与自然和谐相处的典范。保护区北部的天然屏障兴隆岭最高海拔达 3071 米,阻挡了沿汉江河谷北上的暖湿气流和南下的北方寒流,使地处汉中盆地东北部的洋县成为富庶之地,生态优良,物产丰富,气候宜人。长青的崇山峻岭环抱的秀美盆地华阳,是秦岭山中一颗闪亮的明珠,平均海拔 1100 米。著名的华阳古镇就坐落于此,这里曾是傥骆古道上的重要驿站、军事要冲和经济政治重镇。在茂密森林的呵护和众多汇集于此的河流滋养下,环境优美,风调雨顺,物华天宝。人们在享受大自然给予的恩赐时,更加热爱自然、保护自然,使这里成为人与自然和谐的家园。

Changqing, home to the wild Qinling pandas and a natural sanctuary to diverse wildlife, is a showcase of man and nature living in harmony. North of the nature reserve, Xinglongling peaks at 3,071 meters above sea level, forming a natural barrier that prevents the warm, humid air along the Han River valleys from traveling north and barricades the cold northern air from traveling south. This leads to favorable ecological conditions for Yangxian, located northeast of the Hanzhong Basin, giving rise to its rich resources and mild climate. Surrounded by Changqing's many magnificent mountain ranges, the scenic Huayang Basin (average altitude: 1,100 meters) is considered a gem among the Qinling Mountains because the famous, historic Huayang Old Town was built here. Once an important relay stop on the ancient Tangluo Road, this ancient town was also a military hub, an economic and political town. Under the shelter of the dense forest and the nourishment of many rivers converging here, the beautiful environment, timely rains, and the abundant blessed resources allow people to enjoy the gifts of nature. When people fall in love with nature, they enjoy protecting it, making Changqing a harmonious place where man and nature can live in harmony.

野生生命的庇护所
CHANGQING, QINLING
A Natural Shelter to Wildlife

1. 生命之源 Source of life
摄影／赵纳勋 Photography by Zhao Naxun

2. 回归家园 Returning home
摄影／向定乾 Photography by Xiang Dingqian

3. 秋色斑斓 A riot Autumn color
摄影／赵纳勋 Photography by Zhao Naxun

4. 和谐共存 Harmonious co-existence
摄影／向定乾 Photography by Xiang Dingqian

| 1 | 2 |
| 3 | 4 |

野生生命的庇护所
CHANGQING, QINLING
A Natural Shelter to Wildlife

1. 人路亦兽路 Human and animal path
摄影/向定乾 Photography by Xiang Dingqian

2. 朋友相逢 Reunification of friends
摄影/胡万新 Photography by Hu Wanxin

3. 千年古镇 Ancient town
摄影/赵纳勋 Photography by Zhao Naxun

4. 山水华阳 Mountain and water of Huayang
摄影/赵纳勋 Photography by Zhao Naxun

5. 诗画田园 Poetic rurality
摄影/赵纳勋 Photography by Zhao Naxun

6. 相伴相依 Company
摄影/赵纳勋 Photography by Zhao Naxun

7. 秦岭明珠 The Qinling pearl
摄影/向定乾 Photography by Xiang Dingqian

1 信步长青 Strolling in the Changqing Nature Reserve
摄影／向定乾 Photography by Xiang Dingqian

2 朱鹮人家 People and crested ibis
摄影／赵纳勋 Photography by Zhao Naxun

野生生命的庇护所
CHANGQING, QINLING
A Natural Shelter to Wildlife

3. 茱萸人家 People and evodia
摄影／赵纳勋 Photography by Zhao Naxun

4. 鱼米之乡 A land of milk and honey
摄影／赵纳勋 Photography by Zhao Naxun

野生生命的庇护所
CHANGQING, QINLING
A Natural Shelter to Wildlife

秦岭脚下的千年古镇，安静地泊在长青脚下，像画笔任意涂抹出的田园梦乡。
摄影／赵纳勋
The millennium-old town at the foot of the Qinling Mountains nestles silently against the mountains. The landscape brings to mind a pastoral dreamland brought alive by a painter's brush. Photography by Zhao Naxun

索引
Index

动物部分

A
阿穆尔隼 110
暗灰鹃鵙 113
暗绿柳莺 113
暗绿绣眼 113

B
八哥 113
巴黎翠凤蝶 177
白斑翅拟蜡嘴雀 118
白斑迷蛱蝶 176
白背啄木鸟 157
白顶溪鸲 114
白腹[姬]鹟 114
白冠燕尾 114
白冠长尾雉 102
白喉噪鹛 114
白鹡鸰 115
白颊噪鹛 115
白颈鸦 115
白眶鸦雀 116
白领凤鹛 116
白鹭 163
白眉[姬]鹟 116
白琵鹭 162
白头鹎 117
白胸苦恶鸟 163
白腰草鹬 166
白腰文鸟 117
斑背噪鹛 118
斑鸫 118
斑姬啄木鸟 157
斑羚 76
斑头鸺鹠 105
斑胸钩嘴鹛 154
斑缘豆粉蝶 176
宝兴歌鸫 118
豹 72
豹猫 83
北草蜥 84
北方山溪鲵 84
北红尾鸲 118
北椋鸟 118
碧凤蝶 174
冰清绢蝶 176

C
菜粉蝶 176
菜花烙铁头 87
灿福蛱蝶 177
苍鹭 166
苍鹰 110
草兔 81
长尾山椒鸟 119
长纹电蛱蝶 186
橙翅噪鹛 119
橙黄豆粉蝶 177
橙头地鸫 120
橙胸[姬]鹟 121
池鹭 166
赤腹鹰 105
赤颈鸫 122
赤麻鸭 171
赤胸啄木鸟 157
重环蛱蝶 186
翠蓝眼蛱蝶 177
翠青蛇 86

D
大斑啄木鸟 157
大翅绢粉蝶 176
大杜鹃 156, 158
大红蛱蝶 176
大鵟 105
大麻鳽 167
大鲵 84
大沙锥 167
大山雀 121
大卫绢蛱蝶 178
大紫蛱蝶 179
大嘴乌鸦 122
戴菊 130
戴胜 159
黛眼蝶 178
淡绿鵙鹛 120
稻眉眼蝶 178
点斑林鸽 103
雕鸮 107
东方菜粉蝶 179
东方角鸮 106
短趾百灵 122

F
发冠卷尾 120
反嘴鹬 167
方尾鹟 120
斐豹蛱蝶 178
粉红胸鹨 121
凤头鹂鹛 171
凤头麦鸡 167
凤头鹰 107

G
柑橘凤蝶 179
高山短翅莺 122
高山旋木雀 161
戈氏岩鹀 122
骨顶鸡 165
冠纹柳莺 123
冠鱼狗 158

H
豪猪 79
合征姬蛙 90
褐冠山雀 121
褐河乌 125
褐头鹪莺 128
褐头雀鹛 129
黑斑侧褶蛙 91
黑翅长脚鹬 167
黑带黛眼蝶 178
黑短脚鹎 124
黑冠鹃隼 106
黑冠山雀 125
黑喉歌鸲 126
黑喉红尾鸲 124
黑喉石䳭 125
黑脊蛇 86
黑卷尾 125
黑领噪鹛 125
黑眉锦蛇 87
黑弄蝶 179
黑水鸡 167
黑尾蜡嘴雀 124
黑纹粉蝶 178
黑熊 77

236

黑枕黄鹂 129
红白鼯鼠 74
红翅旋壁雀 128
红腹角雉 96, 100
红腹锦鸡 102
红腹山雀 128
红喉歌鸲 131
红基美凤蝶 180
红交嘴雀 129
红隼 117
红头[长尾]山雀 128
红头穗鹛 128
红尾伯劳 126
红尾副鳅 94
红尾水鸲 127
红胁蓝尾鸲 127
红胁绣眼鸟 131
红嘴蓝鹊 130
红嘴鸥 171
红嘴相思鸟 132
红嘴鸦雀 131
虎斑地鸫 130
虎斑颈槽蛇 86
虎纹伯劳 132
花鳉 95
花面狸 81
华西蟾蜍 89
画眉 132
环颈雉 98, 99
鹮嘴鹬 169
黄额鸦雀 134
黄腹山雀 130
黄腹树莺 123
黄钩蛱蝶 180
黄喉貂 74
黄喉鸦 132

黄环蛱蝶 180
黄鹡鸰 133
黄脚渔鸮 104, 107
黄帅蛱蝶 181
黄头鹡鸰 133
黄臀鹎 132
黄鼬 80
黄爪隼 110
灰斑鸠 101
灰背伯劳 134
灰翅鸫 137
灰翅噪鹛 134
灰姑娘绢粉蝶 183
灰冠鹟莺 135
灰鹡鸰 134
灰卷尾 135
灰眶雀鹛 137
灰蓝[姬]鹟 134
灰脸鵟鹰 107
灰椋鸟 139
灰林鸮 134
灰头鸫 137
灰头灰雀 136
灰头绿啄木鸟 159
灰头麦鸡 168
灰头鸦 137
灰头小鼯鼠 76
灰胸竹鸡 102
火斑鸠 101
火冠雀 138

J
矶鸫 168
极北柳莺 140
鲫 94
家燕 138

箭环蝶 180
箭纹粉眼蝶 180
鹪鹩 138
金翅[雀] 139
金雕 108
金凤蝶 181
金眶鸻 169
金眶鹟莺 140
金猫 80
金色林鸲 140
金裳凤蝶 181
金胸雀鹛 140
金腰燕 141
颈槽蛇 86
酒红朱雀 141

K
宽边黄粉蝶 183
宽鳍鱲 94
宽尾凤蝶 183

L
拉氏鲅 95
蓝额红尾鸲 142
蓝翡翠 158
蓝凤蝶 183
蓝喉太阳鸟 143
蓝灰蝶 183
蓝矶鸫 142
蓝尾石龙子 84
蓝鹀 140
酪色绢粉蝶 182
丽纹攀蜥 84
栗背岩鹨 143
栗头鹟莺 144
林麝 75

林猬 78
林鹬 168
领雀嘴鹎 143
领鸺鹠 108
领岩鹨 142
琉璃蛱蝶 182
隆肛蛙 89
绿豹蛱蝶 182
绿背山雀 145
绿翅短脚鹎 145
绿翅鸭 173
绿鹭 165
绿头鸭 172

M
马口鱼 94
毛冠鹿 79
矛纹草鹛 145
煤山雀 145
美线蛱蝶 182
米仓山攀蜥 84
冕柳莺 145
摩来彩灰蝶 184

N
宁眼蝶 184
牛背鹭 168
牛头伯劳 145

P
珀氏长吻松鼠 81
朴喙蝶 184
普通鸬 144
普通翠鸟 159
普通鵟 111
普通鸬鹚 171

普通秋沙鸭 172
普通燕鸥 172
普通秧鸡 168
普通夜鹰 160
普通朱雀 146

Q
强脚树莺 146
翘鼻麻鸭 173
秦岭大熊猫 34-45
秦岭蝮 88
秦岭滑蜥 84
秦岭金丝猴 46-53
秦岭绢粉蝶 184
秦岭羚牛 54-61
秦岭细鳞鲑 92
秦岭雨蛙 90
曲带闪蛱蝶 184
雀鹰 111
鹊鸲 146

S
三宝鸟 160
三道眉草鹀 146
三趾鸦雀 147
山斑鸠 99
山鹡鸰 149
山鹪莺 146
山麻雀 146
扇尾沙锥 164
勺鸡 103
麝凤蝶 183
似鮈 94
寿带鸟 148
树鹨 142
树麻雀 149

双色舟弄蝶 185
双星箭环蝶 185
水鹨 149
丝光椋鸟 146
四声杜鹃 160
松雀鹰 109
松鸦 149
苏门羚 79

T
铜蜓蜥 85
秃鹫 119
突缘麝凤蝶 185

W
网眼蝶 186
乌鸫 148
乌克兰剑凤蝶 186
乌梢蛇 88
巫山角蟾 82
雾驳灰蝶 186

X
喜鹊 151
小鹀 172
小红蛱蝶 186
小灰山椒鸟 151
小麂 74
小鳞胸鹪鹛 150
小太平鸟 151
小鸦 151
小燕尾 150
小云雀 151
斜鳞蛇 88
星头啄木鸟 160
星鸦 151

旋木雀 160
血雉 101

Y
丫纹绢粉蝶 186
崖沙燕 152
烟腹毛脚燕 151
燕雀 112, 152
野猪 78
夜鹭 164
蚁䴕 160
银喉[长尾]山雀 152
银脸[长尾]山雀 152
隐纹花松鼠 76
鹰雕 105
鹰鸮 109
鼬獾 78
玉带凤蝶 187
鸳鸯 172
圆翅钩粉蝶 187

Z
噪鹃 161
藏鼠兔 78
泽鹬 164
窄斑翠凤蝶 187
沼泽山雀 152
赭红尾鸲 128
针尾沙锥 164
直纹稻弄蝶 187
中白鹭 163
中国林蛙 91
中华蟾蜍 89
中华黄蛴蛱蝶 187
中华秋沙鸭 173
中华竹鼠 76

朱鹮 62-71
珠颈斑鸠 99
猪獾 81
紫灰锦蛇 87
紫啸鸫 153
棕背伯劳 153
棕腹柳莺 154
棕褐短翅莺 155
棕颈钩嘴鹛 152
棕脸鹟莺 154
棕头雀鹛 155
棕头鸦雀 154
棕尾褐鹟 154
棕胸岩鹨 154
纵纹腹小鸮 110

植物部分

B
白屈菜 195
百合 213
宝铎草 205
笔龙胆 193

C
苍耳七 218
长柄八仙花 205
齿萼报春 197
川赤芍 211
川续断 217
春兰 195

D
大瓣铁线莲 206
大花糙苏 213
大火草 223

大叶金腰 195
单花无柱兰 207
棣棠 197
点地梅 207
独叶草 188
多花红升麻 209

E
峨眉蔷薇 208

G
甘肃瑞香 197
杠柳 200
狗枣猕猴桃 208
广布红门兰 213

H
荷青花 195
黑蕊猕猴桃 209
红豆杉 223
厚朴 200
忽地笑 217
弧距虾脊兰 213
花锚 221
华北耧斗菜 200
黄花白芨 209
黄毛忍冬 208
黄素馨 200
蕙兰 196

J
剪秋萝 216
金丝桃 214
卷丹 212

K
苦糖果 192

L
辽东丁香 201
列当 201
裂叶地黄 198
领春木 198
流苏虾脊兰 201
轮叶马先蒿 219

M
马桑 199
毛杓兰 209
岷山银莲花 209

N
南山藤 205
拟缺香茶菜 212
扭盔马先蒿 217

P
盘叶忍冬 201
蒲儿根 198

Q
秦岭木姜子 194
瞿麦 212

S
三棱虾脊兰 202
三叶木通 199
山竹花 201
陕西报春 204
扇叶杓兰 205
商陆 215
舌唇兰 204
射干 219
石蒜 221

绶草 214
水金凤 220
水晶兰 221
水杨梅 214
四川杜鹃 199
四照花 201
松潘乌头 218

T
太白杜鹃 198
太白龙胆 218
太白山报春 214
铁筷子 195
头花杜鹃 209
透骨消 202

W
望春玉兰 193
无距耧斗菜 210
五脉绿绒蒿 210

X
藓生马先蒿 203
星叶草 218
秀丽莓 199

Y
延龄草 202
野菊花 223
野鸦椿 222
野樱桃 199
银露梅 211
油点草 215
玉竹 202
鸢尾 202
圆叶鹿蹄草 211

云南大百合 202

Z
中国旌节花 198
中华秋海棠 211
中华绣线梅 205
朱兰状独蒜兰 205
珠根老鹳草 215
梓木草 203
紫斑牡丹 199
紫萼女娄菜 215
紫花地丁 195
紫堇 199
紫罗兰报春 199
紫珠 210
酢浆草 199

秦岭长青

CHANGQING
QINLING